Energies Beyond the State

Energies Beyond the State

Anarchist Political Ecology and the Liberation of Nature

Edited by
Jennifer Mateer, Simon Springer,
Martin Locret-Collet and Maleea Acker

ROWMAN & LITTLEFIELD
Lanham • Boulder • New York • London

Published by Rowman & Littlefield
An imprint of The Rowman & Littlefield Publishing Group, Inc.
4501 Forbes Boulevard, Suite 200, Lanham, Maryland 20706
www.rowman.com

86-90 Paul Street, London EC2A 4NE

British Library Cataloguing in Publication Information Available

Library of Congress Cataloging-in-Publication Data

Names: Mateer, Jennifer, editor. | Springer, Simon, editor. | Locret-Collet, Martin, editor. | Acker, Maleea, 1975- editor.
Title: Energies beyond the state : anarchist political ecology and the liberation of nature / Jennifer Mateer, Simon Springer, Martin Locret-Collet, Maleea Acker.
Description: Lanham : Rowman & Littlefield Publishers, [2021] | Includes bibliographical references and index.
Identifiers: LCCN 2021029567 (print) | LCCN 2021029568 (ebook) | ISBN 9781538159163 (cloth) | ISBN 9781538162187 (paperback) | ISBN 9781538159170 (ebook)
Subjects: LCSH: Political ecology. | Nature–Effect of human beings on. | Human ecology.
Classification: LCC JA75.8 .E554 2021 (print) | LCC JA75.8 (ebook) | DDC 333.7–dc23
LC record available at https://lccn.loc.gov/2021029567
LC ebook record available at https://lccn.loc.gov/2021029568

Contents

Preface

John P. Clark

THE LIBERATION OF NATURE

This work, like the previous two in the series, constitutes a significant advance in the project of developing an anarchist political ecology. As the title indicates, it confronts the central problematic of anarchist political ecology, the liberation of nature, which is not only our inescapable responsibility as wise and compassionate beings but also the only means by which we can liberate ourselves. The project of liberation, that is, the attainment of freedom in its deepest and most expansive sense, is at the same time a process of uncovering and abolishing all forms of domination that stand in the way of that freedom. Accordingly, anarchist political ecology offers a vision of a free society based on mutual aid and solidarity but also works to demythologize all ideologies of hierarchical dualism that underlie the domination of humanity and nature.

As anarchist political ecology develops, it will present a new critical and liberatory reading of geohistory. It will examine the roots of domination in early objectification of nature and of beings in nature. Though it looks to early hunting and gathering societies as examples of cultures founded on communal solidarity and solidarity with nature, rather than on domination, it will also explore the roots of domination that are inherent in the predator–prey relationship, and in the general evolution of the objectifying mind across deep history. Though it recognizes that the toolmaking that was integral to such communal societies did not produce systemic domination, it will examine the ways in which the seeds of the Megamachine and technological domination are present as moments in the long evolution of technique.

The analysis of the domination of nature that is presented in this work shows that it is a complex and deeply dialectical phenomenon. Neither the idea nor the reality of human domination of nature is derived exclusively from the phenomenon of domination of humans by humans. To understand the dialectic of domination, one must look deeply into social history and the history of consciousness, and one must consider the analogical aspects of human knowing, the functioning of the imagination and the work of processes of displacement and transference. Dominating consciousness does indeed project domination within human society on a supposedly 'exterior' natural world. However, at the same time, it projects human domination of that natural world on human social relationships. Similarly, our experience of the free flourishing of wild nature inspires ideas and ideals of freedom in human society, while the quest to liberate the human community inspires hopes for the liberation of the larger natural community from which it is inseparable.

AN ETHICS OF NON-DOMINATION

It is obvious that 'nature', which ultimately includes the whole universe, cannot be entirely dominated by humanity. However, whenever humanity brutally disrupts or destroys the *flourishing* of organisms, populations, species and ecosystems, that is, their attainment of their good, this constitutes real domination of nature. We might even say that it is *nature becoming self-dominating.* The goal of anarchist political ecology is to help reverse this process, so that we can act instead as *nature becoming self-creative and self-liberating.*

Parent points out in chapter 3 that, in view of the enormity of what is at stake, there is a kind of anarchistic categorical imperative to destroy the system of domination, as opposed to merely reforming it. The anarchist critique is based on the insight that, we might say, 'what does not kill domination makes it stronger'. In the words of this author, there is today a 'a globalized predatory system' in which 'the domination of both people and nature' is 'central to the fundamental existence' of that system. It follows that liberation from domination is inseparable from the destruction of the system. As he explains, 'anarchist political ecologists do not seek incremental (i.e. reformist) changes in the current system, but rather, normatively call for the *absolute dismantling* of systems and institutions' of domination. It is significant that he describes this demand as a 'normative' one, and that the destruction of the system is termed an 'absolute' demand. It is one that is both *moral* and *categorical.*

Mainstream political ecology does not have such a clear normative basis. It does have an *implicit*, largely unconscious normative basis in its ideological commitments. These assure that it does not have to examine its own norms and subject them to ruthless critique. Anarchist political ecology, on the contrary, has an *explicit* normative commitment – one that arises from the fact that it is not only political and ecological but also an-archic, that is, opposed to domination. Unlike, for example, Rawlsian liberalism, eco-anarchism has from the outset a 'shared conception of the good.' It has a distinct concept of what is intrinsically good, namely *freedom* and a specific conception of what is intrinsically evil, namely *domination*. Furthermore, it interprets these corelative concepts ecologically, rather than individualistically or anthropocentrically. It is based on an ethical commitment to the abolition of *all forms of domination*, by which is meant the systematic imposition of restrictions on the flourishing and attainment of the good of all beings, and to the realization of *universal freedom*, by which is meant the maximum possible flourishing and attainment of the good for all beings. This process of realization of this good is synonymous with the liberation of nature.

THE WAR ON NATURE

Dunlap and Brock in chapter 5 argue for the need for a political ecology of war and militarism, topics that have been neglected in mainstream political ecology. They are, however, quite central to anarchist political ecology, because *the state* is central to anarchist political ecology. As the anarchist journalist Randolph Bourne famously said, 'War is the Health of the State.' War, whether it be a form of 'internal' or 'foreign' aggression, demonstrates the extreme violence that is latent in the state, and which underlies the everyday, ordinary violence through which the state habitually operates. It is noteworthy that the political concept of 'terror' derives from the state terrorism in the French Revolution. By any objective empirical standard, the state has been the overwhelmingly greatest terrorist force ever since.

Colella points out in chapter 2 the repressed truth that the state is an inherently violent institution. Any purely descriptive analysis of the state recognizes that it is defined by its explicit claim to a monopoly on violence over a defined geographical area. Mainstream political and social theory acknowledges this fact; however, it goes on to treat it as if it were merely an inert theoretical datum. Its all-pervasive violent reverberations throughout the social body are passed over in silence. On the other hand, the project of anarchist political ecology is to show the myriad ways in which war and violence are not only the health of the state but of capital, which acts in close alliance with the modern state (indeed, structurally constituting a capitalist-state apparatus).

Furthermore, as this book demonstrates abundantly, under the dominant world system, industry is the continuation of war by other means (we might even say that conversely war is the continuation of industry by other means). It shows that what is conducive to the health of the state and capital constitutes a plague on the Earth. Even during those brief periods between overt conflict between states, the war waged by capital and the state against the Earth continues unabated, as does the race to develop ever more effective weapons of mass destruction of nature.

One of the great merits of this work is the extensive attention it devotes to exposing extractive industry as the war on nature at its most ferociously ecocidal. Three chapters concentrate explicitly on extractivism and several others touch on the topic. In these discussions, extractive industry is revealed to be the paradigm for the most brutal processes of primitive accumulation or primary appropriation, that is, for the most flagrant and predatory exploitation of the natural world, of workers, and of human communities.

While these chapters devote most attention to mining, Mateer demonstrates in chapter 6 why hydropower is sometimes, with very good reason, classified with the extractive industries. She shows how the massive dam-building projects in India (what Nehru called 'the New Temples of India') have become symbolic of patriarchal, capitalist domination of nature, while at the same time reinforcing class domination by bestowing benefits on the more affluent, and allotting scarcity and displacement to the poor.

THE CRITIQUE OF GREEN CAPITALISM

Another important area of anarchist political ecology that is developed at length in this work is the critique of the 'green economy' and the ideology of green capitalism. Several discussions illustrate versions of the Jevons paradox, which was discussed in another volume in this series. It occurs when the economic and ecological efficiency of production increases, for example, because of 'green' technological innovations and 'green' governmental or corporate policy changes. In such cases, resource use should seemingly decrease, but it instead increases because of increased overall demand. The reason for the 'counter-intuitive' (actually, 'counter-ideological') result is that 'green' initiatives occur within the context and constraints of the larger capitalist economy. It is a problem of structural over-determination. But the evils of green capitalism go far beyond even this serious structural problem.

Various chapters show that supposedly 'green' energy is diverted into very non-green forces of production, and that the operation of 'green' energy systems demands much greater use of non-green energy and other resources than is usually recognized. Colella notes in chapter 2 the many ecologically destructive processes that are involved in the production of solar energy. Dunlap and Brock in chapter 5 observe that polluting companies sometimes create 'sustainable energy' subsidiaries that power and expand their eco-destructive activities. They also note specifically that the devastation of what remains of the Hambach Forest has been justified by appeals to 'green economic recultivation or offsetting initiatives' and by 'corporate social technologies that attempt to marginalize and pacify militant resistance in the area'. And Shannon and Perez-Medina point out in chapter 10 that green energy programs typically turn out to be much more valuable for greenwashing ecologically destructive production than for substantially replacing unsustainable activities. Consequently, we might epitomize such forms of Green Capitalism thus: 'Renewable in means, ecocidal in ends'.

Shannon and Perez-Medina also pose the crucial question of whether the principle of 'grow of die' is truly inherent to capitalism, as is often assumed uncritically. Anarchist political ecology must explore the fundamental issue of whether a 'steady-state' capitalist economy, or more immediately, a steady-state sector within the dominant economy, could emerge under the pressures of ecological necessity. In short, is *growth* the ultimate imperative of capitalism, or is the real imperative maximization of *profit* through the greatest intensification of *economic exploitation*?

It seems plausible that just as capitalism accepts the myth of the free market and free competition, yet embraces state intervention when necessary for its own health and survival, it will adapt to greater or lesser levels of growth (including even maximally exploitative *degrowth*) to the degree that its (albeit ultimately short-term) health and survival depend on it.

RADICALIZING SPACE, PLACE AND SCALE

Political ecology, like geography, sociology, anthropology and other related fields, has hitherto devoted extensive attention to issues of *space*, but considerably less attention to those of *place*. A search for the terms 'spatial' and 'spatiality' will produce an enormous number of references, while searching the terms 'placial' and 'placiality' (or the variations 'platial' and 'platiality') will uncover only a few rare cases of these analogous terms. Whether or not it uses these terms, anarchist political ecology is nevertheless far ahead of most of the field in its exploration of place, the placial and placiality, as this work illustrates at various points.

A spatial issue that it confronts directly is that of borders. Statist thought, which is founded on the idea of sovereign power and authority, is unthinkable without the concept of rigid, clearly defined borders. On the other hand, as Parent points out in chapter 3, anarchist critique questions the very existence of borders in their dominant statist, exclusionary sense. Anarchist political ecology sides with a land-based and Earth-based (i.e. *place*-based) regionalism that recognizes the natural and ecological reality of multiple, shifting, and overlapping boundaries.

The nature of nature is to be always out of bounds. Nature's (non-)boundaries are a refutation on the part of the materially real of the statist fantasy of rigid, unyielding spatial borders. To express this otherwise, all terrestrial phenomena exist in both Cartesian, analytical, quantified, objectified, inorganic, inert, spaces of domination and inhibition, and in situated, historical, living, psychological, cultural and spiritual places of liberation and flourishing. In reality, everything occurs in *splace*, by which is meant the site of the antagonistic dialectic between *space* and *place*. There is always some there, but it is constantly being eroded, or exploded, by the forces of domination. How much of it remains or returns is the result of this dialectic, of this splace war.

This work also contains important reflections on the issue of scale, which has always been a crucially important issue for anarchist thought. It is at the heart of the anarchist tradition's ideas of decentralization, base democracy, self-management and the free commune. Everyone, including anarchist political ecologists, must now face the reality of a world with almost 8 billion people, including 4 to 6 billion already living in urban agglomerations (depending on how these are defined). Some of the analysis here indicates that anarchist political ecology as a radical critique of centralization and domination must confront directly the question of whether urbanization in the age of the megalopolis has become an 'impossible impossibility'. Anarchist thought has begun to investigate the possibility of radical economic degrowth, but we must ask whether it has begun to confront seriously the necessity of radical deurbanization.

GOING BEYOND

Anarchist political ecology embraces a geography of liberated and inhabited places in militant opposition to the official geography (theory) and geoarchy (practice) of dominated and regimented spaces. Accordingly, a major theme of this work, as of the others in this series, is the crucial importance of the indigenous both to survival and to creating a free communal future.

Consequently, any hope for the reversal of the ecocidal course of geohistory will depend on the existence of real-world examples of liberated communities practicing non-dominating mutual aid and solidarity. Zehmisch, in chapter 7, recognizes the groundbreaking work of James Scott, David Graeber and others in rediscovering and disseminating the history of such communities. He cites his own fieldwork with communities that 'continue to live an anarchic way of life until today'. These communities, which the dominant ideology judges impossible, prove indispensable in showing us that 'another world' is not only possible, but has been, and continues to be, actual. We learn to face the real of the culture of death (and of extinction, 'the death of birth') by discovering the real of the culture of life (regeneration and rebirth).

The Introduction to this volume states very clearly the nature of our predicament today: 'On our current trajectory, the extinction of the human species is assured. It is not a question of if, but a question of when.' To the tragedy of such an irreversible destruction of all peoples, communities and cultures must be added the tragedy, in the larger natural world, of the massive destruction of genetic diversity, populations and ecosystems, and the annihilation of millions of species. And as Debney emphasizes in chapter 9, this realization must lead to the creation of forms of large-scale worker and community organization that will be capable of confronting and defeating an entrenched world system of capitalist and statist domination. In short, a lot is riding on a change in 'trajectory'. As the title of this work implies, the fateful question today is whether we can 'go beyond' – whether we have the passion, imagination and not least of all, the practical wisdom, to help unleash those energies that will take us beyond a doomed system of domination to nature's own anarchic order of liberation.

Introduction

The Political Ecology of Resource and Energy Management beyond the State

Jennifer Mateer, Simon Springer and Martin Locret-Collet

We know what a boot looks like
when seen from underneath,
we know the philosophy of boots . . .

Soon we will invade like weeds, everywhere but slowly;
the captive plants will rebel
with us, fences will topple,
brick walls ripple and fall,

there will be no more boots.
Meanwhile we eat dirt
and sleep; we are waiting
Under your feet.
When we say Attack
you will hear nothing
at first

– Margaret Atwood, 1976, p. 193

ANARCHISM, POLITICAL ECOLOGY
AND ENVIRONMENTAL MANAGEMENT
BEYOND THE STATE

Resource and environmental management generally entail an attempt by governing authorities to dominate, reroute and tame the natural flows of water, the growth of forests, manage the populations of non-human bodies and control nature more

1

generally. Often this is done under the mantle of conservation, economic development and sustainable management, but still involves a quest to 'civilize' and control all aspects of nature for a specific purpose. The results of this form of environmental management and governance are many, but by and large, across the globe, it has meant governments construct a specific idea regarding nature and the environment. More often than not, the chosen mode of thinking or what Foucault (1980) would call a 'regime of truth[1]' requires a resource to be destroyed, preserved and/or commodified, all of which constitute different forms of control. These forms of control also extend beyond the natural environment, allowing for particular methods of managing human and non-human populations in order to maintain power and enact sovereignty. Thus, with government-controlled resource management, the physical environment, the ecosystem and the earthlings within it become political objects whose existence turns on the mentalities of the governing authority.

Considering the anthropogenic harm that has befallen the world, including but not limited to mass extinction, the current climate emergency and the 'direct existential threat' faced by all beings on Earth (UN 2018, n.p.), it is clear that the current techniques and technologies of power, control and management are no longer tenable options with which to govern human and more-than-human populations or resources.[2] In order to provide alternative ways of living within our current ecosystems, the field of political ecology offers a significant critique of how governing authorities manage the natural environment (Clark 2012; Perreault et al. 2015). As a line of inquiry, political ecology makes explicit the ways in which governmental regimes of truth surrounding the management and organization of nature have become commonplace or taken for granted, thereby making nature political through how nature intersects with social and economic issues and events (Death 2014; Heynen et al. 2006; Robbins 2004). This tracing of politics and nature generally requires an understanding of power, where conflict over resources is born out of various modes of domination by social elites, private companies and/or the state, or as Latour said, 'Let me put it bluntly: political ecology has nothing to do with nature' (2004: 5), instead, it is about power, control and economic rationales.

Political ecology is a sub-field of critical human geography, focusing on how power is operationalized through environmental conflicts, struggles and management strategies (Bakker 2003; Robbins 2004, 2011; Sultana 2007). In particular, political ecology is a mode of analysis pertinent to understanding the everyday resource-based practices that produce spatial differences and inequalities (Robins 2011; Rocheleau et al. 1996). Common avenues of analysis and examination include who can decide on land use, resource exploitation, 'appropriate,' or legal levels of pollution and which areas of habitat loss are acceptable for the 'benefit' of economic growth (Martinez-Alier 2002; Robbins 2004). In asking these questions, political ecologists examine resource development and extraction as something more than just economic development, but rather how ecosystems and society have become victims of economic mentalities (McKay 2011; Nelson 1995), since as Peet, Robins and Watts (2010: 23) discuss, 'Capitalism and its historical transformations [are] a starting point for any account of the destruction of nature.' Economic development is thus a discourse that allows for the hierarchical ordering of 'productive' systems and entities, entangling the environmental, political and social conflicts (Alimonda 2011). Those people in positions of authority over systems of production (such as people who own capital, governing authorities and groups controlling the state) contribute to the marginalization of certain individuals and subaltern populations (Escobar 1999; Heynen et al. 2006; Bakker 2003; Sultana 2009; Swyngedouw 2004). Political ecologists thus focus on a critique of systems that marginalize certain populations as well as providing an avenue for understanding how lived experiences are a production of various ideologies and power relations that can be traced from the local to the global scale (Bryant and Bailey 1997; Perreault et al. 2015; Robbins 2011; Rocheleau et al. 1996). During the course of these analyses, it often becomes clear that a more localized management structure or community-based management is a viable solution to problems associated with top-down management of natural resources and non-human beings. These bottom-up solutions posed by political ecologists make particular sense given the foundation of political ecology was grown from anarchist roots, relying on management via mutual aid and community cooperation rather than control and sovereignty.

Anarchist geographies have often been incorporated into critiques on and of capitalism, orientalism, social and moral organizing and the state more generally – focusing on how centralized authority and governing inherently organizes humans, other Earth beings, the natural environment and resources hierarchically (Ackelsberg 2005; Crass 1995; Ince 2012; Jeppesen et al. 2014; Morris 2014; Springer 2016). In particular, anarchist Élisée Reclus[3] (1894, 2004, 2013) – one of the forefathers of political ecology – argued for an ecosystem understanding of humans *within* the natural environment, rather than seeing human beings as inherently separate from non-human beings or the biomes within which they live. Reclus, and other anarchists such as Peter Kropotkin, has thus shown many similarities between the critiques, solutions and methodologies present in both political ecology and anarchist geography. For example, Kropotkin (1902) argued that mutual aid was the most effective means of overcoming the struggle between organisms and the environment, or today what we would call environmental conflicts.

Current ideas of nature, the regimes of truth that have in part lead to current resource management practices, give people the illusion of nature 'out there' – a space of rejuvenation, adventure or inspiration[4] (Cronin 1995; Recluse 1905). We see this commonly when people speak of 'finding themselves' in nature, but this understanding of the natural environment encourages an understanding of nature as a neutral space – an area existing outside human history and activities. These lines of thought hold grave consequences for responsible, sustainable living and environmental justice, particularly since nature may be constructed as 'empty,' not as home (Barron 2003; Cronin 1995; Kropotkin 1902).[5] Under this conception of 'empty' nature, human presence in and interaction with nature is a failure – however, as Reclus (1880, 1905) rightly pointed out in the nineteenth century when people stop seeing themselves in nature – as part of an ecosystem – they are more likely to take it for granted, to erase the history of the landscape and to misuse natural resources.

Thus, the natural environment is neither natural nor neutral, but cultural, hegemonic and imbued with power relations – nature does not exist beyond culture and built environments; it is not 'out there' but here, as part of ourselves.

Given the current conceptions of nature, natural resources and society, anarchist political ecology provides not only a critique of current management techniques but also a radical solution to the clear and present danger of the world's new geological epoch – the Anthropocene, which is perhaps more aptly called the Capitalocene (Moore 2016). This book sets out both methods and analyses of case studies demonstrating the management of energies and resources beyond centralized governing authorities – focusing instead on how mutual aid can provide both political emancipation and a more robust sense of environmental security and sustainability. Our aim with these chapters is to advance an 'ecology of freedom', which can critique current anthropocentric environmental destruction, as well as focusing on environmental justice and decentralized ecological governance. While concentrating on these areas of anarchist political ecology, three major themes emerged from the chapters: the legacies of colonialism that continue to echo in current resource management and governance practices, the necessity of overcoming human/nature dualisms for environmental justice and sustainability and, finally, discussions and critiques of extractivism as a governing and economic mentality.

LEGACIES OF COLONIALISM AND ENVIRONMENTAL IMPERIALISM

Modern resource management, overconsumption and the domination of nature and non-human beings is certainly a major cause for the global environmental crisis we find ourselves today. However, modern overconsumption is not the only reason for the sixth mass extinction the world currently faces. We must look to periods of colonialism dating back to the fifteenth century. Further, in many former colonies around the globe, monocultures were set up to benefit colonizers and populations in Europe, wreaking havoc on traditional food systems and ecosystems more broadly (see, e.g., Beinart 2000; Corntassel and Bryce 2011; Kameri-Mbote and Cullet 1997; Moore 2000; Murphy 2009; Studnicki-Gizbert and Schecter 2010; Wood 2015; Ziltener and Kunzler 2013). Political ecologists, such as Escobar (2011), have been strong advocates for the critique of colonialism and neocolonialism within the analyses of modern environmental conflicts. An anarchist political ecology also provides an avenue to understand and critique the legacies of colonialism within current resource management practices. Mateer (this book) adopts this critique in her chapter, discussing how colonialism established the policies for modern unsustainable and inappropriate water management practices. Thomson (this book) also examines the colonial legacies of the G20 Summit meetings, and how the demonstrations against these countries and meetings have included a significant proportion of indigenous activists. Further, Colella (this book) and Dunlap and Brock (this book) discuss the ongoing colonial impacts on

Native Americans in the United States with the extraction of uranium as well as the impacts of wind energy on the peoples of Oaxaca, Mexico – a state with the largest proportion of indigenous peoples in Mexico. Poulados and Lycourghiotis (this book) also provide a concise critique of colonial tools of domination and war by outlining how geographic tools which echo how cartography and other geographic methods were used to control humans and other Earth beings via the formation of borders.

The critiques presented here are not meant to provide a pessimistic vision where one abandons all hope for the future, but rather they seek to provide an impetus to decolonize nature and imagine new 'ways of being that exceed colonial modern ways of living and knowing the world' (Radcliffe 2018: 441), to which we mean validating and recognizing subaltern and indigenous knowledges and understandings of the environment, non-human beings and the place of humans *within* their ecosystems. Instead of the environment as a resource for humans to exploit, the way all actors relate to nature must be understood under a new framework altogether – as culture, home, livelihood and the essence of life itself.

OVERCOMING HUMAN/NATURE DUALISMS

Understanding nature in a new way, as suggested previously, is one of the ways in which our book attempts to critique and overcome the fictitious human/nature dualism. In particular, Maraud and Delay (this book) look at the ways in which resource development exemplifies ideas of modernity and progress in both Sweden and France. The authors describe, in both case studies, that philosophies regarding the environment have shifted from an interdependence on the land and resources to the domination of nature. This shift in mentality has resulted in creating a distance between the local people and their territory. Modernity, then, must be re-conceptualized through an anarchist political ecology lens. Mateer (this book) also picks up this thread of inquiry examining how the relationship between humans and water has been problematized as inauthentic, or inappropriate – that people do not know how to 'use' or 'manage' water correctly. Under this mentality, local communities cannot be trusted to manage or use water sustainably, and thus there is an impetus for governing intervention, particularly from engineering and scientific authorities. Devaluing and even debasing local knowledges is a recurrent theme of colonialism, and in acknowledging the managerial authoritarianism that continues into the present under the guises of development, urban planning and institutional governance, we look to energies that exist beyond such hubris and contempt.

As is common with resource-based interventions, the relationship of humans within an ecosystem has often been based on a constructed dialectic dichotomy, one of the nature versus humanity (Linton 2010; Jonas and Bridge 2003). In the case study presented by Mateer (this book), the dualism within environmental management has fostered an essentialized understanding of water as a scientific abstraction (Linton 2010) and as such the 'truth' about how water and humans interact is understood primarily through a scientific, engineering and economic lens. Zehmisch also critiques the current understandings of and thinking regarding culture/nature juxtapositions, pointing out that there is an interdependence and contextual inseparability of human and more-than-human actors and spheres. For this reason, an anarchist political ecology approach is necessary for conservation discourses in order to challenge anthropocentrism in resource management.

THE HARMS OF EXTRACTIVISM

The third major theme present throughout this book is the critique of extractivism. Extractivism refers to those activities that remove and alter, in large volumes, natural resources before processing. Extractivism includes more than minerals, natural gas and petroleum but also includes agriculture, forestry and fisheries[6] (Acosta 2017) as in the cases discussed by Parent (this book), as well as Dunlap and Brock (this book). In these cases, addressing the exploitation of resources as extractivism helps to explain the devastation to the natural environment and local peoples, but also contributes to understanding the evolution of capitalism. Extractivism is not a new concept but rather began approximately 500 years ago during colonial expansion (Acosta 2013). However, as many of our chapters outline, extractivism did not end after colonies became independent.

In cases of extractivism, the exploitation of raw natural resources takes place regardless of any depletion of the resources, even in their entirety (Acosta 2013). It should come as no surprise then that the benefits, whether economic or otherwise, do not stay within the locality of the resources. For example, Mateer (this book) outlines in her chapter that the vast majority of people living in Himachal Pradesh do not receive the benefits of increased electricity even as their state produces hydroelectric power for the capital of Delhi. However, in traditional discussions on extractivism, the benefits are generally traced to corporations and/or populations in the Global North, making up primary-export accumulation. Thus, as Shannon and Perez-Medina outline in their chapter (this book), extractivism is as predatory as capitalism (Acosta 2013), necessitating the repeated destruction of nature and social life. In order to end this cycle, which has been a constant in the economic, social and political processes of many countries across the world (with varying degrees of intensity), there needs to be an overhaul – a transformation of both our economic systems as well as our relationships within the natural environment (Echeverria 2010). Some scholars have argued for post-extractivist economies, a movement that includes moratoriums and policies that limit resource extraction as well as public campaigns against these projects. These scholars point to the rights to and of nature, as in the Ecuadorian Constitution, as well as the concept of ecological debt (Acosta 2013; Escobar 2006; Hornborg and Martinez-Alier 2016).[7] However, this avenue still reinforces a state-centred approach to resource management and governance. In advocating an anarchist political ecology, this book argues not simply for consultation with local communities but *control* by local communities – not just the inclusion of stakeholders but a total *denouncement* of top-down decision-making.

STRUCTURE OF THE BOOK

In reflecting on the broad implications and themes regarding the legacies of colonialism, extractivism and common human/nature dualisms, this book begins with a discussion on one of the original tools of colonial conquest – maps. The authors Lycourghiotis and Poulados discuss the improved accuracy of digital maps and multi-levelled geographic information systems (GIS) using new satellite and altimetry techniques. The authors then outline how these technologies were originally developed by military organizations to support espionage and navigate new 'smart weapons', however, these uses have rapidly expanded to civilian populations, promising solutions to a number of problems including those related to climate and weather studies as well as the protection of forests and seas. Even with these potentially helpful uses of electronic and satellite cartographic techniques, there are also dangers to both the freedoms and management of all populations of Earth beings, as well as ecosystems more generally. In considering these technologies, the authors outline how these new forms of potential surveillance can be used to protect nature when used without the heavy hand of the state, nor other governing authorities, and are instead in the hands of communities.

The book moves from surveillance technologies to technologies of war and energy in the second chapter. Using an anarchist perspective, author Colella critiques how global powers have sought to use uranium to establish and augment sovereignty via nuclear colonialism (Stoler 2016). These techniques of sovereignty, however, have significant impacts on natural resources and largely impact the people whose lives directly rely on a relationship with their biome. Colella exemplifies these issues by outlining the situation for the Navajo people in the American Southwest. In this case, the author provides an effective critique of capitalism, and how, as an economic model, this strategy necessitates hierarchical structures and imperialism that leads to the development and buildup of nuclear armament, encouraging irresponsible, dangerous and catastrophic uses of nuclear energy. These uses include the control of both human and non-human populations, as well as ecosystems at large. However, in considering the focus of this book, Colella provides evidence of a future for nuclear energy outside of a capitalist economic system through advances in technology and the elimination of hierarchy.

The theme of militarization and the amassing of weapons are also present in the third chapter by Parent, who provides another perspective in the discussion of sovereignty via the technologies of war and environmental destruction. However, instead of focusing on environmental governance and management, Parents chapter discusses the movement of human populations in an age of militarization, technologization of borders and reluctance to recognize climate refugees and other forcefully displaced humans and non-humans.[8] Climate refugee claimants are only bound to increase in the age of the Anthropocene, and thus, this chapter provides a timely call to action. For example, since 2015, the number of displaced persons due to environmental conditions surpassed the number of those displaced due to conflict. The international community, however, has been reluctant to formally acknowledge a new classification of environmental, or climate, refugee. To understand the denial of climate refugees, Parent provides a valuable analysis regarding the interplay of environmental stress and extractivist policies that have been brought about by neoliberal governments. Parent thus stresses how the governments charged with the management of environmental refugees are also responsible for producing environmental refugees. This paradox is unlikely to produce a constructive solution; thus, a radical anarchist political ecology, including Kropotkin's call for mutual aid, becomes an important point of departure for the prevention of societal and environmental ruin.

Our fourth chapter by Maraud and Delay also looks to prevent social and environmental conflicts, providing a tool with which to assist local peoples maintain control of their environments. The tool the authors outline is the TORSO framework, which is used to understand and categorize the attitudes regarding resource management strategies. Using twelve distinct criteria, TORSO identifies various management structures to understand both the trajectories and mentalities that have come to constitute environmental management. In order to fully demonstrate the TORSO method, the authors employ two case studies: farmers of the Roussillon valley in southern France and the Sami of Sápmi, northern Sweden. TORSO is shown to be an important method of emancipation for both populations because of how this tool makes visible the mentalities of governance that have forced communities to abandon traditional resource-based practices as well as the inequalities in decision-making processes between local peoples and a centralized governing authority.

Our book then moves from discussing specific tools and methodological approaches to assist in the emancipation of communities from environmental control to an analysis of how the current ecological crisis has been embedded by (neo)colonial industrialization and the problems of 'green' and sustainable discourses. Dunlap and Brock's chapter, fifth in the book, examines the RWE (Rheinisch-Westfälisches Elektrizitätswerk, the Rhenish-Westphalian Power Plant) a German electric utilities company. RWE is responsible for the Hambach mine, the world's largest opencast lignite coal mine that is slowly destroying large parts of the Hambacher Forst every year that it is operational. Although heavily contested by local populations, the RWE has created a regime of truth that their actions at the Hambach mine are 'sustainable' and can be 'offset', which attempts to both marginalize and pacify militant resistance in the area. The situation in Germany is then compared with wind energy production in Oaxaca, Mexico – one of the greatest wind energy generation sites in the world. The governing authorities of the region, known locally as Istmo, have touted wind energy as a climate change mitigation strategy. However, the traditional indigenous practices of the area are at risk due to the acquisition of land as well as the repression of indigenous dissent to these projects. In this way, the green capitalism of wind energy production in Oaxaca is still part of a normalizing and self-reinforcing extractivist nature of industrial systems. Even renewable energy systems have been used to expand and intensify industrial development and socio-ecological degradation. As such, 'greening' the economy is not enough – a radical change to the economic and social systems is necessary, a change that our authors successfully argue can be achieved through engagement with anarchist political ecology.

Our sixth chapter by Mateer also critiques 'green' energy but takes a slightly different approach by employing an Anarcha-feminist political ecology approach to understanding water management and hydropower in Himachal Pradesh, India. This focus adds explicit attention to how hierarchies are enforced not only through current economic norms but further through patriarchal power structures and the resulting impacts on society. The case study outlined in this chapter shows how climate change has resulted in an economic benefit for the governing authorities of India since with the temperature increases, there has been more glacier melting. The increase of flowing water in the northern state of Himachal Pradesh has provided an economic opportunity for the production of 'green' energy and has thus encouraged the construction of hydropower developments throughout the state. The 'green' economic growth and productivity of Himachal Pradesh has thus slowly been increasing. One of the images used to communicate this economic growth, as a result of hydroelectric power, includes the image of a muscular man embracing the walls of a dyke. This image is indicative of the economic and gendered rationality in Himachal Pradesh that the natural environment is feminized and in need of control through the masculine and capitalist discourses and actions of resource management. Understanding and critiquing this rationality is particularly important for the emerging anarchist political ecology literature since often water bodies and other areas of an ecosystem are often constructed as an 'Other', an object whose only value is based on its utility for human populations, and yet is also constructed as feminine. Both of these constructions are legitimizing the domination of nature.

Chapter 7 continues by similarly examining resource management practices. Author Zehmisch explores the ways in which peasant and indigenous peoples have often evaded state influence and authority within their resource management practices. These resource-based practices and management techniques have allowed for sustainable use, which counters the myth of the tragedy of the commons. In opposition to this, state-based conservation campaigns have actually had negative repercussions on the sustainability of peasant and indigenous resources. As such, there needs to be further support and advocacy of the conditions of anarchy that marginalized communities practise and maintain.

This book continues from the discussion of anarchic communities living in forest spaces, to the potential or urban anarchic spaces. The author of chapter 8 Thomson discusses Kropotkin's theories in relation to the 2017 G20 Summit in Hamburg, Germany. The G20 Heads of State Summits have come to epitomize neoliberal capitalism and industrialization. The industrialization in these G20 countries is so extreme that these nations presently account for 74 per cent of global greenhouse gas emissions, thereby representing both environmental and economic domination, causing environmental and economic injustice and ecocide. Because of what the

G20 Summit represents, these meetings have become sites of anti-capitalist and anti-colonial demonstrations, which Thomson examines in their potential for their enactment of anarchism and radical transnationalism.

The ninth chapter also discusses how to address climate change, but rather than encouraging and discussing cites of revolution, Debney looks at the root causes of climate change and how we might adequately address these causes. The author reminds us that in any discussion around global climate change, we are immediately confronted with multiple challenges around identifying causation while finding solutions that don't merely replicate the thinking and practices that created it in the first place. Debney argues that many contemporary analyses of global climate change reflect the thinking that created it. His chapter outlines this fallacy by focusing on the tendency to perpetuate climate crisis through quick fixes that treat symptoms rather than causes in defence of privilege from change. To that end, Debney examines libertarian socialism as a set of ideas and principles that challenge the historical and social forces that give rise to the climate crisis. He does so by reflecting on Jason W. Moore's work on the Capitalocene, which examines the underlying social relations responsible for global warming by way of a critique of the 'society vs. nature' binary as an enabling ideological pretext. Debney employs Moore's concept of the oikeios, arguing that it provides an opportunity to expand on revolutionary praxis given its commonalities with traditional libertarian socialist notions of workers' self-management of production. From here he contends that the tendency to reproduce the thinking behind the climate crisis in our responses can be overcome.

The tenth and final chapter by Shannon and Perez-Medina continues the critique of capitalism's tendency to destroy the environment as well as encourage socio-economic hierarchies. However, the authors present a different approach with their critique as they argue against catastrophist proclamations that human societies must overcome capitalism or the natural world will be destroyed. Capitalism, particularly neoliberalism, is a very resilient socio-economic system that is able to overcome barriers by incorporating and coopting challenges that seem, at first glance, to arise from insurmountable obstacles. The authors then effectively argue that anti-capitalism must be rooted in an ethical opposition to capitalism, which is possible with a radical anarchist political ecology that can provide solutions to the constant metamorphosis of capitalist economic and ecological impacts.

NOTES

1. In *Power/Knowledge* (1980), Foucault discusses a regime of truth as a product of the society within which a discourse is decided as true, arguing,

> truth is a thing of this world: it is produced only by virtue of multiple forms of constraint. And it induces regular effects of power. Each society has its own regime of truth, its 'general politics' of truth: that is, the type of discourse which it accepts and makes function as true. (131)

2. Discussions around the term 'resource' have pointed to the potential problem with the term – that a 'resource' is one that is inherently to be used by humans. Although the term

is imperfect, we do use it throughout this book while also recognizing that resources water, forests, air and so on all have importance outside of human use.

3. Reclus did not use the term political ecology to describe his work, but we consider his scholarship as an antecedent to this area of inquiry (Purchase 1997).

4. These ideas of nature have shifted overtime, as previously nature was often considered harsh, and even a place of despair especially if you consider the 'wilderness' of the Judeo-Christian bible (Cronin 1995).

5. This is a very European construct as in many cultures, nature as described here is a foreign idea. For example, the terms used for nature, the environment or wilderness might instead be referred to as the pantry, kitchen, home or yard (Dowie 2009; Newbery 2012). Further, other scholars describe nature and the land 'as a system of reciprocal relations and obligations' (Coulthard, in Hallenbeck et al. 2016: 112), or as a 'web of interdependent relations' (Radcliffe 2018: 442).

6. Agriculture is considered extractivist because generally the growing for export is mono-cultured and thus destructive to biodiversity and local food security.

7. Ecological debt is defined by Escobar as 'countries or social groups that appropriate biomass in excess of their biological production, or that pollute beyond their capacity to process their pollutants incur in an ecological debt with those who bear the burden of it' (2006: 10).

8. Animals are often forgotten in discussion of climate refugees, however, recently media attention has brought their unique struggles to the forefront. For example, in India a wild tiger was found in the bedroom of a family home during a monsoon that had flooded the home of the tiger (Sottile and Suri 2019). The animal was considered calm, and did not pose any danger, but was merely trying to migrate to higher and dryer ground. Other animals have been known to do the same including deer, rhinos and elephants (Sottile and Suri 2019).

REFERENCES

Ackelsberg, M. A. 2005. *Free Women of Spain: Anarchism and the Struggle for the Emancipation of Women.* Oakland: AK Press.

Acosta, A. 2013. Extractivism and neoextractivism: Two sides of the same curse. *Beyond Development*, 61: 61–86.

———. 2017. Post-extractivism: From discourse to practice—Reflections for action. In Carbonnier, G., Campodónico, H., & Vázquez, S. (Eds.), *Alternative Pathways to Sustainable Development: Lessons from Latin America* (pp. 77–102). Geneva: Graduate Institute of International and Development Studies.

Alimonda, H. (Ed.). 2011. *La Naturaleza Colonizada: Ecología política y minería en América Latina [Colonized Nature: Political Ecology and Mining in Latin America].* Buenos Aires: CLACSO.

Atwood, M. 1976. Song of the worms. In *Selected Poems: 1965–1975*. Boston: Houghton Mifflin Company.

Bakker, K. 2003. A political ecology of water privatization. *Studies in Political Economy*, 70: 35–58.

Barron, J. 2003. Innu support and the myth of wilderness. *Pathways: The Ontario Journal of Outdoor Education*, 15(1): 4–8.

Beinart, W. 2000. African history and environmental history. *African Affairs*, 99(395): 269–302.

Bookchin, M. 1971. *Post-Scarcity Anarchism*. Berkeley: Ramparts Press.

Bourdieu, P. 1989. Social space and symbolic power. *Sociological Theory*, 7(1): 14–25.

Bryant, R. L., & Bailey, S. 1997. *Third World Political Ecology*. New York: Routledge.

Clark, J. P. 2012. Political ecology. In *Encyclopedia of Applied Ethics* (2nd ed., Vol. 3, pp. 505–516). San Diego: Academic Press.

Corntassel, J., & Bryce, C. 2011. Practicing sustainable self-determination: Indigenous approaches to cultural restoration and revitalization. *The Brown Journal of World Affairs*, 18: 151.

Crass, C. 1995. Towards a non-violent society: A position paper on anarchism, social change and Food Not Bombs. *The Anarchist Library*.

Cronon, W. 1995. The trouble with wilderness; Or, getting back to the wrong nature. In Cronon, W. (Ed.), *Uncommon Ground: Toward Reinventing Nature* (pp. 69–90). New York: WW Norton and Company.

Death, C. (Ed.). 2014. *Critical Environmental Politics*. New York: Routledge.

Dowie, M. 2009. *Conservation Refugees: The Hundred-Year Conflict Between Global Conservation and Native Peoples*. Cambridge, MA: The MIT Press.

Echeverría, B. 2010. *Modernidad y blanquitud*. México: Editorial ERA.

Escobar, A. 1999. After nature: Steps to an anti-essentialist political ecology. *Current Anthropology*, 40(1): 1–30.

———. 2006. Difference and conflict in the struggle over natural resources: A political ecology framework. *Development*, 49(3): 6–13.

———. 2011. *Encountering Development: The Making and Unmaking of the Third World*. Princeton, NJ: Princeton University Press.

Foucault, M. 1980. *Power/Knowledge*. New York: Pantheon.

Goodrich, J., Lynam, A., Miquelle, D., Wibisono, H., Kawanishi, K., Pattanavibool, A., Htun, S., Tempa, T., Karki, J., Jhala, Y., & Karanth, U. 2015. *Panthera Tigris*. The IUCN Red List of Threatened Species 2015: e.T15955A50659951. DOI: 10.2305/IUCN.UK .2015-2.RLTS.T15955A50659951.en.

Haig-Brown, C., & Dannenmann, K. 2002. A pedagogy of the land: Dreams of respectful relations. *McGill Journal of Education*, 37(3): 451–468.

Hallenbeck, J., Krebs, M., Goonewardena, K., Kipfer, S. A., Pasternak, S., & Coulthard, S. E. 2016. Book review forum: Red skin, white masks: Rejecting the colonial politics of recognition. *The AAG Review of Books*. DOI: 10.1080/2325548X.2016.1146013.

Heynen, N., Perkins, H. A., & Roy, P. 2006. The political ecology of uneven urban green space the impact of political economy on race and ethnicity in producing environmental inequality in Milwaukee. *Urban Affairs Review*, 42(1): 3–25.

Hornborg, A., & Martinez-Alier, J. 2016. Ecologically unequal exchange and ecological debt. *Journal of Political Ecology*, 23(1): 328–333.

Ince, A. 2012. In the shell of the old: Anarchist geographies of territorialisation. *Antipode*, 44(5): 1645–1666.

Jeppesen, S., Kruzynski, A., Sarrasin, R., & Breton, É. 2014. The anarchist commons. *Ephemera*, 14(4): 879.

Jonas, A. E., & Bridge, G. 2003. Governing nature: The re-regulation of resources, land-use planning, and nature conservation. *Social Science Quarterly*, 84(4): 958–962.

Kameri-Mbote, P., & Cullet, P. 1997. Law, colonialism and environmental management in Africa. *Review of European Community International Environmental Law*, 6(1): 23.

Latour, B. 2004. *Politics of Nature: How to Bring Sciences into Democracy*. Cambridge: Harvard University Press.

Linton, J. 2010. *What Is Water? The History of a Modern Abstraction*. Vancouver: University of British Columbia Press.

Martinez-Alier, J. 2002. *The Environmentalism of the Poor: A Study of Ecological Conflicts and Valuation*. Cheltenham: Edward Elgar Publishers.

Moore, J. 2000. Sugar and the expansion of the early modern world-economy: Commodity frontiers, ecological transformations, and industrialization. *Fernand Braudel Center*, 409–433.

Moore, J. 2016. *Anthropocene or Capitalocene? Nature, History, and the Crisis of Capitalism*. Oakland: PM Press.

Morris, B. 2014. *Anthropology, Ecology, and Anarchism: A Brian Morris Reader*. Oakland: PM Press.

Murphy, J. (2009). Environment and imperialism: Why colonialism still matters. *Sustainability Research Institute*, 20: 1–27.

Nelson, R. 1995. Sustainability, efficiency, and god: Economic values and the sustainability debate. *Annual Review of Ecology and Systematics*, 26: 135–154.

Newbery, L. 2012. Canoe pedagogy and colonial history: Exploring contested space of outdoor environmental education. *Canadian Journal of Environmental Education (CJEE)*, 17: 30–45.

Nixon, R. 2011. *Slow Violence and the Environmentalism of the Poor*. Cambridge, MA: Harvard University Press.

Peet, R. 2007. *Geography of Power: Making Global Economic Policy*. London: Zed Books.

Peet, R., Robbins, P., & Watts M. (Eds.). 2010. *Global Political Ecology*. London: Routledge.

Perreault, T., Bridge, G., & McCarthy, J. (Eds.). 2015. *Routledge Handbook of Political Ecology*. London: Routledge.

Pickering, K. 2016. Moral languages in climate politics. In Hayward, C., & Roser, D (Eds.), *Climate Justice in a Non-Ideal World* (pp. 255–276). Oxford: Oxford University Press.

Radcliffe, S. A. 2018. Geography and indigeneity II: Critical geographies of indigenous bodily politics. *Progress in Human Geography*, 42(3): 436–445.

Reclus, E. 1880. *Histoire d'une montagne [History of a Mountain]*. Paris: Hetzel.

———. 1894. *The Earth and Its Inhabitants: Universal Geography*. London: J. S. Virtue.

———. 1905. *L'Homme et la Terre* (Vol. V) [*Man and Earth* (Vol. 5)]. Paris: Librairie Universelle.

———. 2004. *Anarchy, Geography, Modernity: The Radical Social Thought of Elisée Reclus*. Lanham, MD: Lexington Books.

———. 2013. *Anarchy, Geography, Modernity: Selected writings of Elisée Reclus*. Oakland, CA: PM Press.

Robbins, P. 2004. *Political Ecology: A Critical Introduction*. London: Blackwell Publishing.

———. 2011. *Political Ecology: A Critical Introduction* (Vol. 16). Hoboken, NJ: John Wiley & Sons.

Rocheleau, D., Thomas-Slayter, B., & Wangari, E. (Eds.). 1996. *Feminist Political Ecology: Global Issues and Local Experiences*. London: Routledge.

Sottile, Z., & Suri, M. (2019, July 19). Tiger Takes Catnap on Bed in Indian Home After Fleeing Huge Floods. *CNN World*. https://www.cnn.com/2019/07/19/india/india-tiger-bed-flooding-intl-hnk/index.html.

Springer, S. 2016. *The Anarchist Roots of Geography: Towards Spatial Emancipation*. Minnesota: University of Minnesota Press.

Stoler, A. L. 2016. *Duress: Imperial Durabilities in Our Times*. Durham, NC: Duke University Press.

Studnicki-Gizbert, D., & Schecter, D. 2010. The environmental dynamics of a colonial fuel-rush: Silver mining and deforestation in New Spain, 1522 to 1810. *Environmental History*, 15(1): 94–119.

Sultana, F. 2007. *Suffering for Water, Suffering from Water: Political Ecologies of Arsenic, Water and Development in Bangladesh*. Minneapolis: University of Minnesota Press.

———. 2009. Fluid lives: Subjectivities, gender and water in rural Bangladesh. *Gender, Place and Culture*, 16(4): 427–444.

Swyngedouw, E. 2004. *Flows of Power: The Political Ecology of Water and Urbanisation in Ecuador*. Oxford: Oxford University Press.

United Nations, General Assembly, Address 73: Report of the Secretary-General, A/63/332 (25 September 2018). https://www.un.org/sg/en/content/sg/speeches/2018-09-25/address-73rd-general-assembly.

Ziltener, P., & Kunzler, D. 2013. Impacts of colonialism. *American Sociological Association*, 290–311.

1

Panoptic Geography

Humans and Nature under Surveillance

Sotiris Lycourghiotis and George Poulados

In recent decades, we have witnessed the rapid growth of satellite and geographic technologies. A series of new satellite and air (altimetric) techniques have emerged radically improving the accuracy of digital maps and enabling the development of multi-level geographic information systems (GIS). Initially, these techniques were developed by militaries and security organizations for espionage and the navigation of the new 'smart weapons'. However, these technologies have rapidly expanded to use by civilians, promising new research fields and technical solutions for a range of problems such as ship, plane and car navigation; traffic control; town planning studies; climate and weather studies; forest protection; protection of the sea; disease prevention and control; and innumerable others. This chapter explores the fundamental question of human autonomy and nature in the light of new electronic and satellite cartography techniques (GIS, GNSS, etc.).

An in-depth presentation of the new geolocation techniques and GIS will highlight the dangers that their possession might constitute. Taking these circumstances into consideration, is there space to develop independent social and ecological action within this framework? Can the new forms of panoptic cartography be used to protect nature and humans? What ethical and political questions does the possession of mega data from governments and global organizations pose? Contrary to current practice, the free dissemination of geographic information can trigger a new critique of dominion policies on the planet. What arguments does the modern image of the world give us through the eyes of the satellites that can advocate for an ecological, egalitarian, self-managed humanity?

In this chapter we will attempt a historical overview of the technological innovations of geography, we will try to analyse the military and non-military uses of satellite geography, and we will reflect on the political and ethical questions posed by the new era of geolocation. Is there a different perspective for the future? Are there any answers for the future from an anarchist, ecological perspective?

HISTORICAL BACKGROUND

On 4 October 1957, the USSR launched the first satellite Sputnik into Earth's orbit. Although Sputnik was carrying only a simple radio transmitter, it caused a delirium of excitement in the countries in the East, and at the same time an unprecedented freeze in the West. The reason for these dichotomous reactions was obvious. The ability to put a technical object in orbit around the Earth provided unlimited ability to monitor one's opponent, while at the same time, making it theoretically possible to launch an attack from space. Efforts in the West to fill the gap in technologies with the USSR inaugurated the new satellite or 'space wars' era.

The space race continued with the construction of satellites for two different uses. The first purpose was the aim of monitoring people and the environment using photographic techniques and, the second was for the purpose of creating a global navigation system. The former had a clear military purpose for the United States and the USSR – the rivals sought to gain precise knowledge of the opponent's geospatial, their terrain and morphology, military and transport facilities and so on. Since ancient times, this geophysical knowledge has the most important information of war, and that is why wars have triggered the art and development of cartography. The use of orthophotographic techniques (Thrower and Jensen 1976), that is, taking multiple images of the landscape from different angles during the course of the satellite, has enabled the creation of three-dimensional imaging with the use of satellite images. In the past, the creation of three-dimensional maps and ground models was made possible only by multiple terrestrial measurements using the traditional instruments of topography. The use of orthophotographic techniques is based on basic geometry concepts and can be understood by the way human eyes function. Since humans have two eyes, we are afforded a better understanding of the three-dimensional shape of an object. Nowadays, the technique of orthophotography is known to us by Google Earth (Patterson 2007). The three-dimensional images that we are provided with are the result of this technique. Along with orthophotography techniques, an attempt has been made to construct altimetric satellites (Smith 1997) that could, by means of an electromagnetic wave, calculate, for each position of the satellite, its exact distance from the ground. Thus, satellites could create three-dimensional maps from simple two-dimensional photos. The basic principle of the altimeter (McGill et al. 2013) used by modern planes is based on the simple use of the gravimeter. As the atmospheric pressure decreases linearly with respect to the height from the ground surface, a barometer measuring atmospheric pressure can calculate the height. However, this could not help the satellites much because, on the one hand, the height to be measured would be the barometric height, that is the height relative to the surface of the sea, and on the other hand, the accuracy of such

an instrument was not satisfactory. Thus, an effort was made to use distance measurement technology with electromagnetic waves (electronic distance measurement
– EDM) in order to calculate the distance of the satellite from the ground surface
or the surface of the sea. The basic function of the EDM (Rüeger 2012) is based
on the measurement of the time needed for an electromagnetic signal to return to a
satellite, once sent from it, travelling through the atmosphere and reflecting on the
surface of the Earth. If we know precisely the time needed for the signal to make
this route and divide it by two and multiply it by the speed of light, we have the
distance from the satellite to the ground surface. This measurement is presented with
several problems. On the one hand, high-precision clocks are required, and on the
other hand, the velocity of the electromagnetic waves when travelling through the
atmosphere decreases according to the climatic conditions and, therefore, must also
be taken into account when correcting the final result. Altimetric satellites with substantial accuracy became a reality only in the early 1990s with the TOPEX/Poseidon
(1992), Jason-1 (2001), and Jason-2 (2008) NASA satellites (Masters et al. 2012).

We have already mentioned that the second objective of the space era and the
arms race of the Cold War was to create a global navigation system. The fundamental
aim of such a system was, initially, the precise navigation of ships and transatlantic
submarines, as well as the guidance of ballistic missiles in the future. It is no accident
that the US Navy made these basic efforts because until the Second World War, the
calculation of the position of a ship at sea was made with the traditional use of the
astrolabe. The navigator would have to go to the deck, measure the angular distance
of some stars or the moon, and then use his or her clock and some tables to figure
out their position. This knowledge that had emerged from the attempts of the royal
astronomers of the British Empire in previous centuries contributed to the world
domination of the British Empire. However, for the needs of the modern era, these
techniques became obsolete since its accuracy was small, and it was very dangerous,
especially for submarines that had to emerge to find their position constantly. During the Cold War, submarines had become the primary weapon of the first assault,
since they could carry nuclear-powered missiles into the opponent's backyard without being noticed.

During the Second World War, the British Navy had developed terrestrial radio
navigation systems, such as the LORAN (long-range navigation) and the Decca
Navigator System, based upon the operation of land-based stations of a known
position emitting radio waves (Samaddar 1979). By calculating the distance of at
least two or three such stations, the navigator of the vessel at sea was able to locate
his position accurately, by means of an intersection. However, the use of this radar-
like system was limited to distances relatively close to the shores and could not be
extended to oceans due to the curvature of the Earth. In advancing this technology, an important step was made due to an almost random incident. To prove that
the launch of Sputnik was true, the Soviet Union claimed that anyone could hear
the transmitter broadcasting from the satellite anywhere on Earth. Two American
physicists at Johns Hopkins University working in the Applied Physics Laboratory,
William Guier and George Weiffenbach, began capturing Sputnik's signal. They

soon realized that because of the Doppler effect (Dicke 1953) they could locate the position of the satellite. The following spring, the two physicists explored and solved the inverse problem: how to locate the user's position, if the satellite has a given position. By solving this problem, a series of satellites were launched leading to the well-known Global Positioning System (GPS) (Schenewerk 2003). Overcoming the limitations of the previous systems of the 1960s, the GPS had twenty-four satellites in 1973. The basic principle of GPS is simple to capture. Based on the dispersion of the satellites, the terrestrial user can have visibility of at least four satellites at a time. So, if one can figure out the location of a handheld receiver from four satellites, for which one knows his/her position, then one can accurately calculate his/her position using a geometric intersection in space. It is the same principle applied by a terrestrial system, such as the LORAN. It is the same principle of triangulation applied by a topographer from antiquity to the present day. The only difference is that our position is not calculated as the intersection of two or three circles but of four spheres. Thus, one can find the location of a place not only on a plane but also in three-dimensional space. With the GPS, the world's first three-dimensional geo-referencing system was launched. To understand how we track our position with GPS and other satellite techniques, we can think of the way dolphins and other animals track the position of objects in space through delayed sounds, essentially using a sonar system that is similar to the ones we humans have on ships. Animals use sound waves while satellites make use of electromagnetic waves.

Today we know the GPS because of its usage policies, which we will discuss further in this chapter. Only a few of its capabilities are accessible by the public, as many of the system's capabilities are exclusively for military use. The Soviet Union did not stay out of this race. In 1976, the country initiated its own system, also known as GLONASS, and since the early 1980s there has been a massive launch of satellites. GLONASS is based on the same principles as the GPS, and its basic difference is that it does not cover all regions of the globe with the same accuracy, as the main concern of the Soviet Union, and then the Russian Federation, was to cover its country. Other states have developed similar technologies, which for users provides the opportunity to combine systems and technologies so the user of a receiver can have access to information from many satellites of different systems and thus improve his accuracy. The current combination of these systems is called Global Navigation Satellite System (GNSS) and includes GPS (USA), GLONASS (Russia), Galileo (European Union), QZSS (Japan) and BeiDou (China).

POLITICAL AND MILITARY USE OF SATELLITE GEOGRAPHY IN OUR TIMES

After the end of the Cold War, there was a gradual release of the navigational systems we discussed previously for more general use such as for research and politics. As a result, there has been a development of a series of applications that we can see in our everyday lives. The release of these technologies combined with the development of third industrial revolution applications, satellite geography technology has become accessible to a range of applications, in fields that no one until now had imagined. The combination of geographic techniques with the internet, and through it with mobile phones and social media, has created hundreds of new approaches that have substantially changed the field of modern geography, but at the same time created several ethical questions for geographers in the field of anarchist studies and political ecology. At the same time, military applications have not declined in number, but on the contrary have expanded in civilian–military fields such as preventive repression and political control. Also, the number of strictly military applications of unmanned aircraft and 'smart weapons' has increased. All these have created a new era that is interesting to examine in detail.

During the Cold War, a major focus for both the United States and the Soviet Union was the acquisition of the precise knowledge and mapping of the geospace. This goal was not only about the geometric form of cartography, that is, the creation of high accuracy maps, but also to gain qualitative information. Thus, what is sought is not only the precise knowledge of the three dimensions of the landscape of the opponent but also where facilities are located, what they contain, how many people are in an area, what jobs they do and so on. The underlying logic of these levels of information is also the basic logic of most modern geographic applications used by civilians, which is reduced, among other things, to the concept of the GIS. However, before we begin this discussion it is important to provide a brief overview of how GIS as a technology has been used.

The release of a GPS frequency for civilian use by the US government gave a huge boost to geographic and topographical research. Even though the US Army techni-cally limited the precision provided by GPS for civilian use, a series of applications have been developed. The first and foremost application had to do with navigation. Airplanes, ships, in the air and at sea, vehicles and travellers (such as climbers) on land have been able to calculate their position with significant accuracy. Significant bugs and difficulties in calculating a position with the traditional methods have become virtually obsolete. For example, the fishermen of the open seas could now mark the exact coordinates of one good catch and thus improve their chances of catching even more fish while small boats without a radar could navigate safely near the shores under challenging visibility conditions or at the ocean. Those responsible for removing millions of tons of debris from World Trade Center after the 9/11 terrorist attacks had GPS receivers placed on the trucks to direct drivers so as not to disturb the traffic flow of the city, although the job lasted several months.

A second set of applications is related to classical topography. The low accuracy of GPS measurements, which ranged from a few centimetres to tens of metres, was overridden by using two techniques. The first had to do with the number of measure-ments, a technique of precision improvement known since antiquity. A coordinate can be defined with greater accuracy, the more times it is measured, since the errors follow the normal distribution (Taylor 1997). The second technique has to do with the use of a second identical and stable GPS receiver in an area near the first receiver.

Since the uncertainty of GPS measurements comes from 'noise' that is added from the Earth's atmosphere, two identical receivers that record simultaneously at almost the same point, and thus receive a signal from the same satellites, will have approximately the same noise. However, if one of the two is at a fixed point, then it will only record 'noise'. Thus, with the new technique we can remove this noise from the original receiver and 'clear' the signal increasing. The accuracy for GPS was achieved and therefore suitable for classical topography.

The main advantage of satellite topography over terrestrial areas is that the former no longer needs to reduce the points of a local measurement to a reference system, local or global. Using satellites, the position is immediately defined in a global datum/global geodetic reference system (Grafared and Okeke 1998). The mapping of a road no longer requires complicated computations for engineers, but only a GPS receiver. At the same time, monitoring natural phenomena, such as landslides, became much easier. The GPS also revolutionized geophysical research. Earth tectonic movements, the movements caused by an earthquake or the 'bloating' of a volcano were challenging to study before the introduction of the GPS since it was not possible to determine a fixed point that was not affected by the geophysical phenomenon, and from that point to measure the relative movement of the rest. There was no unaffected point of reference. However, with a satellite system, the reference system is outside the surface, and so the movements of the Earth can be calculated with precision. Thus, the movements that are induced by an earthquake on the surface of the Earth, the form of the deformation of an inverse rupture that causes a tsunami or the detection of the magma's movement in a volcano could now be studied (Puglisi and Bonforte 2004; Blewitt et al. 2006). The various systems recording ocean waves using floating GPS beacons (Watson et al. 2008) can be considered as another important application, whose data can be used to predict and highlight dangerous conditions for navigation, provide instructions for the routes of the ships and to record tsunami waves, as they are generated, and to give time for coastal evacuation through a tsunami early warning system (TEWS). With the gradual improvement in the accuracy and frequency of sampling of the GPS, steady progress has been made on the phenomenon of oscillation and small movements (Psimoulis and Stiros 2008), such as the winding of a bridge, the oscillation of a skyscraper. Also, the phenomenon of an earthquake could be completely recorded for the first time. These applications directly link geolocational techniques with ecology and can help in preventing disasters and protecting the natural environment and humans.

A third set of applications has to do with the interconnection of the information provided by the GNSS (GPS) with internet applications, mobile phones and social media. The changes that have occurred over the past decade in this category of applications have a direct and rapid impact on our lives and give us the feeling of a technological revolution. The applications are many, and therefore we will only be able to discuss some examples. Car traffic in urban areas is still one of the most important problems, which has been quite difficult to study. The reason is that it was impossible for researchers to place devices and sensors that would measure traffic load, traffic density and speed at many points in an urban network. Therefore, the studies were time-consuming, costly and focused only on certain points. However, with the use of GNSS in mobile phones today, a Google traffic app can identify the key traffic sizes at every point of an urban network at any given time (Zhou 2014) and thus inform its users of the route time and traffic status. Every moving mobile phone is converted, without the user knowing it, into an independent traffic researcher. The utilization of this information from an integrated Intelligent Transport System (ITS) can guide urban traffic at all times and overcome severe traffic congestion (Barceló et al. 2005). An extension of these applications is the various bus and taxi management systems operating in conjunction with mobile phone applications facilitating the movement of people and services in cities, as well as dozens of applications being used for the management of transport of all modes (air, sea, rail), whether passenger or freight. Similarly, some applications can record and can accurately predict traffic and user habits through the recording of their purchases. Thus, in conjunction with the data from various social media, profiles of both consumer and wider social behaviour can be created. It is important here to emphasize that such applications are not interested in full registration among all users, as even data concerning a small percentage of the population (up to 5 per cent) can give an accurate picture of the whole population. At an experimental level, there are several models of predicting not only the consumer but also social–political behaviour. If such apps proliferate, they may raise many ethical questions about using them, as we will discuss subsequently. For example, if an authoritarian power seeks to control and foresee every movement of its citizens, and through this prediction guide behaviour, so that citizens think they are free to make decisions, while in reality they are being guided, we will have an Orwellian state. We must pay more attention to these dangers and mobilize people so as to prevent the development of such phenomena.

An app which is quite interesting has to do with the use of these technologies in blocking the development of epidemics (Albert et al. 2003). By recording the path of the last days of people suffering from an epidemic disease, when they are identified as ill (e.g., in a hospital), a geographic view of the paths of many patients is formed, so that common starting points of the disease can be identified. These points are quarantined, and the spread of the disease is confined.

Unmanned robotic vehicles, helicopters and airplanes are a category that combines GNSS with data transfer over the internet, and a whole class of applications has been created in recent years. In these applications, we can organize transport and product classification either within logistics or between the company and the buyer. Amazon is considered as a pioneer in these applications and conducts significant research into both the robotic organization of its supplies and the creation of postal services using drones (Bamburry 2015). Another important category of these systems is the possibility of live geographic recording and tracking of the terrain from above. For example, forest maps can easily be created, burning fires can rapidly be traced, future disasters can be foreseen, forest deforestations and other environmental disasters can be identified (Fedra 1999).

On the other hand, many repression and control applications have been identified. By using multiple geo-referenced data (i.e., information referring to a particular geographic location) deriving from monitoring drones, the internet, traditional archival data or from the social media, police authorities create predictive combinatorial algorithms and can construct maps of potential crime. So, they can focus their repression on specific locations (Chainey and Ratcliffe 2013). Despite their relative accuracy, we observe that these systems have a major problem, they tend to confirm the 'biases' of their data. For instance, in some US cities, these systems have shown a greater possibility of crimes being committed in areas where people of colour live, leading to an even more selective racist policing, resulting in community reaction. With the use of thermal cameras for drones, border surveillance systems and recording of irregular migratory flows have also advanced.

Another application that has to do with repression is the GNSS system bracelets applicable to suspects and prisoners who are on leave (Patil 2014). These systems can operate indoors even where there is no satellite visibility, since they locate the position of a person with the combined use of a satellite system and the signal from mobile phone antennas. Thus, when such a receiver is in an enclosed space, its geographical position can be determined with relative accuracy from the known coordinates of the locations of the mobile antennas from which it receives a signal. By the same reasoning, the intersection of the cycles we analysed earlier for the GPS, the geographical position can be spotted with this signal as well. This application has been used on young children in Great Britain (due to abductions), and a new application on watches measures many biometric characteristics of humans which can be applied to almost all population groups, generating new ethical dilemmas (Michael et al. 2006).

Before discussing the political and ethical dilemmas of the new age of geo-referencing, it is important not to forget the many military applications that are being created. The best-known application is that of the 'smart weapons'. In the past two decades, we have seen them appear in the wars against terrorism in Iraq, Afghanistan and so on. It has been an unprecedented upgrading of the ballistic navigation systems. With the use of a GNSS receiver and an accelerometer with three degrees of freedom, a missile can permanently correct its position in the air and hit its target with very high accuracy (Kaplan 2006). With the combined use of unmanned tracking aircraft and the use of spy satellites that provide live images of very high precision of less than 1 metre, the conduct of the war by the armies holding this technology becomes almost virtual, as they can hit targets with great accuracy from a long distance, without being involved locally, and at the same time, watch live the result of their actions. With military access to a plethora of data from e-mails and social media (whose companies provide access to the army), targeted surveillance and strikes against suspects are also possible in any part of the world. Many troops throughout the world (United States, Russia, Great Britain, etc.) are currently developing robotic unmanned armored vehicles that will be able to conduct wars on battlefields without soldiers. The basic idea of all the military applications is to reduce the risk of human cost for the army that owns these applications.

ETHICAL AND POLITICAL QUESTIONS IN THE ERA OF GEO-LOCALIZATION (GEO-REFERENCE)

The examples outlined in this chapter support the view that we have entered a new age of geo-referencing. More and more information, more and more humans, movements, actions and thoughts are recorded. With the use of mobile phones and social media on a daily basis, each person produces a number of megabytes of geo-referenced information. The data generated and accumulated, the so-called mega data, increases at immense rates, creating new multi-level geography. This complex new condition has generated a series of ethical and political issues.

The use of heterogeneous geographic information provided by users today can help a number of applications solve major technical problems of cities such as traffic. However, the question is: who owns and who manages the mega data? To date, their management is done by giant monopolies such as Google or Facebook. These companies own data volumes whose economic wealth is valued at tens of billions of dollars while their potential political power may be ten times as much. Or, as a newspaper had written a few years back, 'data is the oil of the future'.[1] Thus, the determination of the US government to create an access agreement to the data of those companies for its federal agencies cannot be considered as an accident. The protection of the users' private data – who they are, where they are, what they do, how they work, with whom they communicate and what they say – is guaranteed only against other users and not the state. Similar 'indiscretion' has recently been demonstrated by the Chinese government when it decided that it would have access to the data of the users of a social media application. The use of mega data by governments and large monopolies does not only imply the end of privacy but can potentially have detrimental effects. In a global community where only Facebook users are 2.5 billion (2017 figures) and the total interconnected users of all networks that collect geo-referenced data are more than half the world's population, the possession of data that has to do with location, traffic, preferences, privacy, consumer habits, beliefs and so on by a single power raises many questions. Knowing the profile, movement and behaviour of each person create the ability to control, define and guide behaviour to the fullest extent. This knowledge may, in the future, create a model of governance that may well exceed Orwell's literary projections. Such knowledge will be identified with power, just as Michel Foucault predicted. In such an environment, the notion of human freedom should be re-examined.

The concept that we would like to introduce here is that of *the geo-referenced existence*. This concept consists of a new form of the subject, for which its position (in its first stage) and its actions (to a lesser extent) are known by the visible or invisible panoptic eye of authority. Each position at a known time creates a condition of objectification of the subject, which is closely associated with a reification process. Thus, the subject-person is gradually transformed into an object and a thing. Here, not only do we have the classical Marxian concept of alienation, the person alienated by the product of his work, but we have an even deeper and qualitatively superior reification.

For the industrial worker, what becomes the object of reification is the body, with the objectification of existence, we have a complete instrumentalization of all aspects of a person, even his thoughts. Consensus about the techniques of the new geography takes place in silence precisely because either we can no longer live without the technical facilities of the new applications, or because these facilitates upgrade several practical aspects of our everyday life. The degree of accuracy and time coverage of the position, the range of control and recording of movement ultimately determine the degree of objectification – reification and can be considered a measure inversely proportional to human freedom. The geo-referenced information systems in their full promise are the 'utopia' of every dominating system.

With the new panoptic geography, we can talk about an epistemological super-field, a constructive unifying sphere, seeking to bring together all levels of empirical observation in a single language. This super-field launched by geo-referenced information systems is based on the reduction of heterogeneous information (both qualitative and quantitative, dynamic and static) to a unique reference system. Why is the reduction in such a system? In our opinion, there is a substantial analogy with the market economy. As the market seeks to fully monetize all things, whether material goods and services or other values, for their inclusion through money in the law of supply and demand, so does the new geography seek a full geo-referenced common equivalent. If capitalism perceives everything as an economy, as monetary values that can be bought and sold, power sees everything as geography. The reduction of all subjects to objects strips them from all their qualities and destroys the infinitude of the meanings of each position to a bare objectification. Further if for humans, this objectification coincides with the practical purposes of market life, for nature and the environment this market-based objectification poses greater risks. The reduction of each physical parameter to a geographical and quantitative dimension is the other aspect of the commercialization of nature, or if you prefer, the necessary step for it. It is precisely this reduction that the anarchist worldview is fighting against, seeing every being, not only animals and humans but also nature as a subject and never as an object or a thing. The worldview of anarchism is the one that attempts to reverse this perspective.

WHAT IF WE REVERSED THIS PERSPECTIVE?

Let us think for a while how things would be from a different perspective. What would the image of the world be like if the mega data of the geographic systems were not in the hands of the world's rulers, but free for every one of us to use? There is, of course, a critical parameter that we should never ignore: there is no such thing as a neutral technology. A scientific idea may bear the scientist's good intentions, but a technological implementation is always linked to the specific purposes of society. As long as we live in market societies, our technologies are geared towards its goals. A prospect of a free exchange that would not accumulate information in the hands of the powerful could potentially transform the goals of modern geography. From a geography of control and of the market, we could move to critical geography that would reveal the social inequalities as a result of the capitalist mode of production and would bridge the world on the basis of equality and solidarity. We all live on the same planet, and we all share the same natural resources, so a different geography could focus on such an orientation. The modern geography of the world that the satellites and the various information systems reveal to us could, if we read it in the light of another perspective, dissolve many of the reactionary ideas, such as the idea that there are pure races, or that borders are natural phenomena. It can also reveal to us that defending the environment and nature will either be the work of a united and egalitarian humanity, or competition will fatally destroy the environment and people.

CONCLUSION

Knowledge of geospatial has always been the prerequisite for domination. From the ancient empires to the Greek merchants and settlers, from the Romans to the Venetians, from the colonial powers to the modern West, the forms of war and domination may have changed, but the primary parameter has not: the privileged knowledge of space. The eye from the perspective of which most information was and has been known is almost always the eye of the master(s). If the Panopticon that Bentham (1791) envisioned about prison surveillance tends, nowadays, to encompass society as a whole, we ought to, along with the practical benefits of the new era, highlight the dangers that such a perspective may bring. The meaning of freedom can change over the centuries, but some parameters remain unchanged. And perhaps the most basic one is that of self-determination since liberty requires a degree of indeterminacy of the subject in relation to the eye of the authority. If every such indeterminacy, if every autonomy is lost, then the future of freedom is dismal.

In the seventeenth century, Descartes formulated the idea of the self-reflecting subject, Cogito. Since then, this concept has defined Western thought. Human beings as a subject 'should always be treated as an end in themselves and not as a means to something else' claims Kant in his third formulation of the categorical imperative (2017). At present, it has become clear that the notion of the self-reflecting and the self-determining subject is being challenged is under attack from everywhere, seeking to make it disappear from the face of the Earth. Perhaps a modern anarchist geography should first and foremost preserve this subject.

NOTES

1. This phrase was recently used by Mohammad Al-Gergawi. Its origin, however, is unknown

REFERENCES

Albert, D. P., Gesler, W. M., & Levergood, B. (Eds.). (2003). *Spatial Analysis, GIS and Remote Sensing: Applications in the Health Sciences.* CRC Press.

Bamburry, D. (2015). Drones: Designed for product delivery. *Design Management Review, 26*(1), 40–48.

Barceló, J., Codina, E., Casas, J., Ferrer, J. L., & García, D. (2005). Microscopic traffic simulation: A tool for the design, analysis and evaluation of intelligent transport systems. *Journal of Intelligent and Robotic Systems, 41*(2–3), 173–203.

Bentham, J. (1791). *Panopticon or the Inspection House* (Vol. 2). London: Mews-Gate.

Blewitt, G., Kreemer, C., Hammond, W. C., Plag, H. P., Stein, S., & Okal, E. (2006). Rapid determination of earthquake magnitude using GPS for tsunami warning systems. *Geophysical Research Letters, 33*(11).

Chainey, S., & Ratcliffe, J. (2013). *GIS and Crime Mapping.* John Wiley & Sons.

Dicke, R. H. (1953). The effect of collisions upon the Doppler width of spectral lines. *Physical Review, 89*(2), 472.

Fedra, K. (1999). Urban environmental management: Monitoring, GIS, and modeling. *Computers, Environment and Urban Systems, 23*(6), 443–457.

Grafarend, E., & Okeke, F. (1998). Transformation of conformal coordinates of type Mercator from a global datum (WGS 84) to a local datum (Regional, national). *Marine Geodesy, 21*(3), 169–180.

Kant, I. (2017). *Kant: The Metaphysics of Morals.* Cambridge University Press.

Kaplan, C. (2006). Precision targets: GPS and the militarization of US consumer identity. *American Quarterly, 58*(3), 693–714.

Masters, D., Nerem, R. S., Choe, C., Leuliette, E., Beckley, B., White, N., & Ablain, M. (2012). Comparison of global mean sea level time series from TOPEX/Poseidon, Jason-1, and Jason-2. *Marine Geodesy, 35*(suppl 1), 20–41.

McGill, M., Markus, T., Scott, V. S., & Neumann, T. (2013). The multiple altimeter beam experimental Lidar (MABEL): An airborne simulator for the ICESat-2 mission. *Journal of Atmospheric and Oceanic Technology, 30*(2), 345–352.

Michael, K., McNamee, A., & Michael, M. G. (2006, June). The emerging ethics of humancentric GPS tracking and monitoring. In *International Conference on Mobile Business, ICMB'06* (pp. 34–34). IEEE.

Patterson, T. C. (2007). Google Earth as a (not just) geography education tool. *Journal of Geography, 106*(4), 145–152.

Patil, P. B., Chapalkar, S., Dhamne, N. D., & Patel, N. M. (2014). Monitoring system for prisoner with GPS using wireless sensor network (WSN). *International Journal of Computer Applications, 91*(13), 28–31.

Puglisi, G., & Bonforte, A. (2004). Dynamics of Mount Etna Volcano inferred from static and kinematic GPS measurements. *Journal of Geophysical Research: Solid Earth, 109*(B11).

Rüeger, J. M. (2012). *Electronic Distance Measurement: An Introduction.* New York: Springer Science & Business Media.

Samaddar, S. N. (1979). The theory of Loran-C ground wave propagation—A review. *Navigation, 26*(3), 173–187.

Schenewerk, M. (2003). A brief review of basic GPS orbit interpolation strategies. *GPS Solutions, 6*(4), 265–267.

Smith, L. C. (1997). Satellite remote sensing of river inundation area, stage, and discharge: A review. *Hydrological Processes*, *11*(10), 1427–1439.

Taylor, J. (1997). *Introduction to Error Analysis, the Study of Uncertainties in Physical Measurements*. University Science Books.

Thrower, N. J., & Jensen, J. R. (1976). The orthophoto and orthophotomap: Characteristics, development and application. *The American Cartographer*, *3*(1), 39–56.

Watson, C., Coleman, R., & Handsworth, R. (2008). Coastal tide gauge calibration: a case study at Macquarie Island using GPS buoy techniques. *Journal of Coastal Research*, *24*(4), 1071–1079.

Zhou, L. (2014). *U.S. Patent No. 8,626,439*. U.S. Patent and Trademark Office.

2

Uranium

Capitalism, Colonialism and Ecology

Chris Colella

Uranium is a radioactive element with ninety-two protons, and isotopes ranging from uranium-233 to uranium-238 (Hammond 2000). Uranium is a mineral
of great importance and interest in our generation, especially to those who wish to
analyse, understand and especially resist the status quo. Its history is intertwined
with energy and ecology, with warfare and weaponry and with capitalism and
colonialism. If we want to understand the many crises our world is facing today,
and if we wish to come up with a proper solution to them, we must understand
uranium.

There are a couple of terms that are necessary to define. First 'colonialism', which
under my definition refers to the control by a group of people over a different land,
people and/or culture. Although more nuanced definitions certainly exist, this simplified definition suits my endeavour here.

Several different methods of colonization exist and have existed throughout the
history. The first, and the one most people are probably familiar with, is 'settler colonialism'. This is where a particular group sends people, usually of the lower classes of
society, to settle in a particular land area, claiming that land for the colonizers and
pushing out or exterminating the native peoples. The second form of colonialism is
'exploitation colonialism', in which the goal is to exploit the natural resources of a
colony for the economic gain of the colonizing power, but not so much to settle in
the area and fill it with the colonizers' people. A reason for this type of colonization
is that, in certain situations, it may be easier to use the native population as a source
of cheap labour or slave labour for the extraction of resources or production of commodities. The third form of colonials, which will be addressed in much greater detail
later in this chapter, is 'internal colonialism'. Internal colonialism is when certain
regions or peoples within a country experience severe inequity relative to the rest of

the country, due to a systemic and/or deliberate uneven development of the 'core' and 'periphery' of the country.

The second major term necessary for definition is 'capitalism'. There are many different definitions of capitalism, particularly those discussions on what makes it different from other modes of production. However, for the purpose of this chapter, the most accurate and succinct description of it is that capitalism has three elements that make it *capitalism*. These three elements are private ownership of capital, the selling of goods and services in a market and the system of wage labour for the non-owning masses. These rules provide an incomplete description of every aspect of capitalism; however, they provide a good groundwork for describing and examining the implications of capitalism.Capitalism has been the dominant mode of production in the world for a few hundred years, and the dominant mode of production for the Western world in particular for some time longer. Naturally, laws governing land and natural resources in the world are therefore influenced largely by (and created specifically to serve) capitalism, which in itself has grown out of Western culture and ideas. Laws regarding natural resources tend to have a lot of emphasis on 'individual rights', which tend to be highly valued by Western society (McHarg 2010). However, capitalism has the effect of causing the privatization of land and resources to expand faster than ever before, due to the high rewards yielded by those who own, and the positive feedback loop caused by individuals claiming proprietary rights over what was previously the commons, thereby lessening the ability of non-owners to support themselves, creating more demand for property. As previously public goods (e.g., land, energy, minerals) become further privatized, or as new human needs develop as society and technology develop (such as the need for internet as computers and communication become more and more necessary in daily life), 'property' becomes an even more powerful concept. Furthermore, under capitalism,[1] resources are not distributed based on need but rather based on access to capital, increasing populations and wealth will create more demand and cause scarcity (ibid.). However, as with any economic or political system in which power is distributed based on property and ownership thereof, with a limited amount of property in the world, distribution of wealth is guaranteed to be unequal (ibid.). This situation then grants more power to those that own land, allowing them to generate more wealth and then acquire more property, thereby consolidating more property (and therefore power) into the hands of the few and away from the many. With the concentration of wealth and power into the hands of a few, and such drastic inequality of wealth over the population, it is inevitable that this wealthy minority will use their power to exploit the people, land and resources of the world to benefit themselves, at the expense of whoever happens to fall victim of those actions, typically and especially those at the bottom of the dominant social hierarchy.

URANIUM MINING

Uranium is a radioactive element; Uranium-238 has a half-life approximate to the age of the earth, and only emits alpha radiation. Uranium-235, on the other hand, has a half-life about one-sixth of the half-life of U-238, and emits both alpha and gamma radiations. Still, a lump of pure uranium would emit less radiation than a lump of granite. Though this is dangerous, the real danger of uranium comes from two sources other than uranium's own radioactivity: from the chemical toxicity of uranium (comparable to that of lead) and from the radioactivity of elements that uranium decays into over time. Uranium decays into the element radium, which is far more radioactive, and generally when geologists search for uranium, they actually look for radium, which serves as a decent indicator of uranium (World Nuclear Association n.d.).

Uranium was first found in the sixteenth century by silver miners in St. Joachimsthal in what is now the Czech Republic. It was dark and greasy, and stuck to miner's picks, and typically stained surrounding rocks green, orange and yellow. The minors referred to this substance as 'pitchblende' and 'uraninite'. As the mining went on, and the uranium was cast aside as waste, miners were struck with a disease called 'bergkrankheit', or 'mountain disease'. Many explanations for the disease were hypothesized, but nobody thought that it involved the weird, velvety black rocks they were finding. Over a century later, in 1789, samples of the mysterious rocks ended up in the hands of a Berlin pharmacist, Martin Klaproth, who managed to separate the uranium from the pitchblende. He found it had a vibrant yellow colour to it, which made it good for use as a dye and for colouring glasses. He decided, rather than naming it after himself (as was custom at the time), he would name it after the newly discovered planet, Uranus (Zoellner 2009).

Centuries later, in the Dutch Congo, King Leopold largely ignored the possibility of mining, instead preferring to extract the much cheaper-to-produce rubber from the region. However, mining companies discovered something that Leopold had not realized: under the ground in the Congo, very close to the surface, were vast quantities of bismuth, zinc, cobalt and tin. At the time, the peoples of the Congo were paid roughly the equivalent of 20 cents a day to mine the ore. Workers were not allowed to select their own occupations, and taxes were kept impossibly high, which created a form of debt slavery that forced workers to toil their entire lives in the mines for little-to-no reward. Patches of high-grade uranium were found at a mining site called Shinkolobwe in 1915, and although uranium was considered mostly useless, the presence of uranium meant the presence of radium, the most valuable substance on Earth at the time. A gram of radium could be worth about $175,000, which was 30,000 times the price of gold. The reason for such a high value was that war was an incredibly profitable business to be in at the time, and the United States was already a frequent customer of the Belgian Congo, being the world's largest user of Congo cobalt, for aircraft engines. Shinkolobwe would go on in the 1940s to provide roughly two-thirds of the uranium used in the atomic bombs dropped on Hiroshima and Nagasaki (ibid.).

There are three general methods for the extraction of uranium from the Earth. Shallow orebodies are removed via open-cut mining, where miners dig up a large, open pit to extract the ore and remove waste rock and overburden (overlaying rock). Deeper orebodies are generally removed through underground mining, where access shafts and tunnels are dug to reach the deeper ores. This method generally removes less waste rock and has less of an environmental impact than that of open-cut mining. The third way is through in situ leaching (ISL), also called in 'situ recovery' (ISR), which uses a leaching solution that dissolves the uranium trapped in porous, unconsolidated materials, such as gravel or sand, in groundwater and pumps it out of the ground. If done properly it has the least environmental impact of any form of uranium mining; however, if it is done improperly it can damage underground water supplies. ISL is not permitted where it may threaten potable water supplies, though this is not to say that it does not happen. The good side of ISL is that there is no major disturbance of the ground or removal of waste rock, as the uranium solution is pumped directly out of the ground to a treatment facility. Most uranium production in the United States and Kazakhstan is done by this method, and, in 2013, 47 per cent of the world's uranium production came from ISL, an immense increase from 16 per cent in 2000. In the United States, ISL is seen as the most cost-effective and environmentally stable way of extracting uranium, and if it is done safely and properly, ISL leaves behind only several boreholes as damage, and rehabilitation of the land is far simpler than with open-cut and underground mining (World Nuclear Association, n.d.).

Conventional uranium mines will have a mill where the ore is crushed, ground and leached in tanks with sulphuric acid to dissolve the uranium oxide. The solution is processed to separate the uranium from the rest. The remaining ore is rocks and dust referred to as 'tailings', which is most of the original ore and contains most of the original radioactivity of the ore. One of the minerals in these tailings is radium, which through the radioactive decay process emits radon gas, which is highly radioactive. Because of this, measures must be taken to limit the radon gas emissions from tailings. These measures include covering the tailings in layers of clay and topsoil and eventually planting a layer of vegetation over it. Radon occurs in most rocks and in trace amounts in the air we breathe, but in higher concentrations it poses a serious risk to human health (ibid.).

In 2014, the Organization for Economic Co-operation and Development (OECD) Nuclear Energy Agency stated in a report titled 'Managing Environmental and Health Impacts of Uranium Mining':

> Uranium mining and milling has evolved significantly over the years. By comparing currently leading approaches with outdated practices, this report demonstrates how uranium mining can be conducted in a way that protects workers, the public and the environment. Innovative, modern mining practices combined with strictly-enforced regulatory standards are geared towards avoiding past mistakes committed primarily during the early history of the industry when maximising uranium production was the principal operating consideration. (n.p.)

Although I do believe in safer, more ecological safe methods and practices in mining, including that of uranium, I am highly sceptical of the ability of hierarchical, centralized models of organizations such as capitalism and the state to carry out such methods in a way that benefits humanity. Most people in positions of power are

removed from the consequences of their actions, or at least the social and environmental consequences; in a centralized, bureaucratic and hierarchical organization,
those with the power to make decisions are only able to see costs and benefits in
terms of finance and statistics, never in terms of vital resources and human beings.
Without decentralized, participatory planning and execution of plans, it is almost
certain that somebody is going to get the short end of the stick, which is especially
worrying in situations involving human lives and the health of the planet. For these
reasons, the management of uranium would be best informed by an anarchist model,
which would provide the best possible ethical and ecological outcomes.

From the 1930s to the 1970s, Navajo people, Hopi people and Mormon people
were used by mining companies to mine for uranium in the four corners region of
the United States (Arizona, New Mexico, Utah and Colorado). The miners, who
were largely indigenous, were not informed of the dangers of working in such close
proximity to uranium without proper protection by the mining companies nor by
the US government. Uranium tailings were left uncovered and unmarked all over
the region near the mines. The leftover ore was used to build houses for the miners
and their families in the area. The Atomic Energy Commission (AEC) was aware of
the dangers posed to uranium miners prior to uranium mining due to the studies
of uranium miners in Europe that had previously taken place, and throughout the
1950s, the commission oversaw uranium mining in the region without informing
miners of the dangers of uranium (McLeod 1983). During this period, the Navajo
people mined about 4 million tonnes of uranium, largely used by the US government in the production of nuclear weapons during the Cold War, and there are now
259 abandoned uranium mines in the state of New Mexico alone (Frosch 2009).

Ever since the uranium mining in the four corners region and the Navajo nation,
uranium has contaminated the water that the peoples of the region use for drinking, cleaning and cooking. The people that work and live near old uranium mines,
tailings and downstream from them end up consuming high amounts of uranium
throughout their life, which causes sickness and early death among them. It should
be noted that the children of dead parents in these situations often are taken by the
government and put in the foster care system, robbing the children of their families
and cultural identities (Spitz 2009). In addition to contaminating the drinking
water, the uranium contaminated the every brick that people's homes were built out
of, and living in one of the uranium houses could be up to the equivalent of receiving
553 chest X-rays per year (McLeod 1983). The radiation in the region dramatically
increased the rates of birth defects among infants in the Navajo nation as well.

In 1997, the United States Environmental Protection Agency finally measured the
levels of radiation in Monument valley and discovered that many water sources had
been contaminated due to the waste being piled up along the Colorado River, just as
the indigenous peoples had been saying for years, and that in some areas the radiation was around eighty times the federally designated dose limit for human beings
(Spitz 2009). The 2000 documentary *The Return of Navajo Boy* was finally able to
assist in triggering a federal investigation into uranium houses in the four corners
region. The federal government has even promised reparations to all those affected

by uranium mining, though they often will refuse to help lung cancer patients due to the use of traditional tobacco (ibid.). In the year 2008, the US EPA finally made a five-year plan to clean up all Cold War uranium contamination on Navajo land, though vast areas of land and many families have still been left out of this plan. Federal and tribal scientists tested radioactivity over the 27,000 square miles of the reservation, searching for the lung-cancer-causing mineral, radium (Frosch 2009).

In a personal account, Bertha Nez, a resident of the southeastern edge of the Navajo nation, said that 'I'm sandwiched between these tailings piles. The kids have asthma, my sister had cancer – lymphoma – then my dad, and some people that worked in the uranium mine, they have respiratory problems and some have kidney problems'. 'There were a lot of things people weren't told about the plight of Navajos and uranium mining' (Stephen B. Etsitty, executive director of the Navajo Nation Environmental Protection Agency) said, 'These legacy issues are impacting generations. At some point people are saying, 'It's got to end' (Jung 2013).

This internal colonialism by the United States was used during the Cold War for multiple purposes. The first and foremost was the development of a large and sophisticated nuclear arsenal for use in intimidating the rest of the world into submission for (neo)colonialism and imperialism abroad. The second was the deliberate destruction of native homelands in an attempt to force the indigenous peoples of the region to either assimilate into the culture of the United States, or to leave and no longer be a bother to the white colonizers. The third was an attempt by the state and capitalism to further reduce the health and population of the indigenous peoples of North America by inflicting the horrible diseases and conditions that come along with living in uranium mining waste. And yet these practices are on the rise again in recent years, as several multinational corporations have begun to mine in Mount Taylor, a site considered sacred to the Navajo people and many other surrounding peoples, referred to as *Tsoodził* by the Navajo people. The project is called the Roca Honda Mine and it is the largest mine in the nation. It has the potential to pollute local groundwater sources and destroy culturally significant sites, but it is also estimated to create 2,400 jobs and over $1 billion in economic activity, an offer hard to refuse in a state that ranks third in poverty. However, the mine is predicted to affect up to 70 acres of land designated by the forest service to be a traditional cultural property (ibid.).

Investors are beginning to take notice of uranium mining lately, despite uranium mining was never really recovering from a crash in the early 1980s, likely due to the recent surge in desire for alternative energy sources. The project has been taken over by the Canadian-based company Energy Fuels, and, using the underground mining method, they hope to extract 1,000 tonnes of uranium every day for 9 years. Jackie Jefferson, a neighbour of Bertha Nez, has said about the mine, 'We just don't want no other uranium stuff coming to us. We don't want to be a dumping ground' (ibid.)

Despite the negative stigma around uranium, the world continues mining it to use for the production of energy and, in a more sinister vein, weapons. Over two-thirds of the world's production of uranium comes from Kazakhstan, Canada and

Australia, with a rapidly increasing amount extracted using ISL. Uranium production overall has also been increasing in recent years, rising from 41,282 tonnes in 2007 to 60,496 tonnes in 2015; Kazakhstan's production has increased by approximately 300 per cent since 2007; Canada's has increased by about 50 per cent; and while Australia has actually decreased production by about a third. The United State's uranium production has remained about the same over the past decade, but it would not be surprising if it were to begin rising again. Currently, over half of uranium mining is handled by state-owned mining companies, which generally prioritize secure supplies of uranium over actual market considerations. Uranium production from mines took a serious downturn in the late 1980s and early 1990s but has been steadily on the rise ever since. Recently, in 2012, uranium supply hit the highest it had been since 1988, and over the next few years could easily surpass its peak in 1980. The largest producing mine in 2015 was McArthur River mine in Canada (7,354 tonnes of uranium), followed by Cigar Lake, Canada (4,345 tonnes of uranium) and Tortkuduk & Myunkum, Kazakhstan (4,109 tonnes of uranium). The United States has approximately 207,400 tonnes of known recoverable uranium resources, 4 per cent of the world's known uranium resources (World Nuclear Association, n.d.).

The environmental aspects of uranium mining are largely the same as other metalliferous mining, especially that of heavy metals which are generally highly toxic, but the radioactivity of uranium and other related ores is what makes uranium mining especially worrisome compared to other forms of mining. ISL runs the extra risk of polluting groundwater if done improperly. Mining equipment used for uranium mining generally cannot be sold once it is being used due to exposure to radiation, so it has to be buried along with the rest of the waste. Uranium mining sites must get prior approval from the governing bodies with jurisdiction over the land before being used, they must comply with all environmental standards of those governing bodies and they are subject to international standards and external audits. This is not to say that it is not possible to ignore regulations, or that the regulations are commensurate to the risk posed to workers, communities and the environment, but there are at least some general standards of safety that prevent disaster, and the regulations are almost always getting better (ibid.). One could hope that a solution would be to control over industry and energy production in a communal, democratic manner so that the people most affected by any mining or milling may make the decisions as to whether and how they want to operate these facilities, as people with decision-making power would be those affected, and could weigh the benefits and costs (which are social, personal and environmental, not simply monetary as viewed by central organizers such as businesses and governments).

An environmental engineer working for the New Mexico Mining & Minerals Division (MMD), James Smith, said, 'Total mines that have been really well cleaned up? Ummm. One. You can't see uranium. You don't know that it's there. So, it's really hard to be able to recognize what to do.' Cleaning up uranium mining sites may sound easy in theory, but in practice it is very difficult to determine exactly what the impact of the mine has been, and to what extent it is necessary to cover it up.

Covering old uranium mines can range from piling rocks and topsoil over it to filling an entire mine shaft with concrete, it really depends on the particular site. Juan Velasquez, senior vice president of environmental affairs at Strathmore Minerals, one of the companies involved with the Roca Honda Mine project, easily admits that industry has had bad practices in the past, but also insists that things have changed and that they know better now. The New Mexico Mining Act of 1993 introduces very strict regulations on mining in the state, and they have more regulations requiring companies to clean up any damage they do to the environment. Eric Jantz, a lawyer with the New Mexico Environmental Law Center, said that 'the single biggest problem environmentally is the water situation', he continued, 'In order to start mining, the company is going to have to de-water the mine, which means they're going to have to pump out millions of gallons of groundwater from the mine area' (Jung 2013). Especially with ISL, there is extra risk of releasing heavy metals into local water sources. 'For the communities most immediately impacted, the benefits are negligible and the costs are huge', Jantz said, and for the ones with the power to carry out the deeds, the costs are negligible and the benefits are huge (ibid.).

NUCLEAR WEAPONS

The story of the capacity for uranium to produce high quantities of energy, which would be used both in nuclear weapons and in nuclear power, begins with the discovery of the neutron. In 1932, English physicist James Chadwick discovered the neutron, a particle theorized by Ernest Rutherford a decade earlier. Neutrons have no charge (as opposed to the +1 charge of the proton and −1 charge of the electron), and can only be detected by the way they cause other particles around them to move. Rutherford would comment soon after the discovery that anyone who thought that neutron collisions could provide any sort of useful power was 'talking moonshine' (Zoellner 2009). But perhaps the story starts earlier with a book written in 1914 by socialist and science-fiction author H. G. Wells: *The World Set Free*. The book described a mineral that could use radioactivity to set off a chain reaction to liberate the binding energy of atoms and create an explosion; destruction on a global scale would then ensue. Leo Szilard, another physicist, was annoyed by Rutherford's comments and, inspired by what he feared after having read Wells' book, decided that he needed to make sure that this idea of such a weapon never became public. Szilard patented the idea in the name of the British Admiralty and it was promptly forgotten. Not long after this, a new form of uranium, U-235, would be discovered by physicists at the University of Chicago, a much rarer form that could be used for nuclear fission (ibid.).

In September of 1938, Adolf Hitler annexed the Sudetenland, a region of Czechoslovakia with an ethnic German majority, and consequently gained control of the mining town of St. Joachimsthal, one of the world's only known sources of uranium at the time. Fears grew that if Germany were to gain control of Belgium, then they would have control over Shinkolobwe, and the United States and Britain would have little-to-no access to uranium supplies, while thousands of tonnes would be necessary to crush, separate and enrich to get enough U-235 to create a uranium bomb. Luckily (though unsure at the time), the likelihood of the Germans developing a nuclear bomb was slim due to Hitler's distrust of technology, and his dismissal of the uranium bomb as the 'Spawn of Jewish pseudo-science' (ibid.). The US government on the other hand decided to go ahead with the project and ordered the purchase of a depopulated piece of land in Washington state near the Columbia River. This became the Hanford Site, where the government would secretly manufacture plutonium, and it would turn into one of the most polluted sites on the Earth. George B. Kistiakowsky, present for the Trinity explosion, the first nuclear weapon ever detonated, described the explosion as 'the nearest thing to Doomsday that one could possibly imagine. I am sure that at the end of the world – in the last millisecond – the last man will see what we have just seen' (ibid.).

Nuclear weapons were developed during Second World War by the United States under the Manhattan Project and were designed to serve one purpose and one purpose only: to destroy cities and population centres. A common trend of US military tactics is the use of what was once referred to as 'irregular warfare', attacking cities, villages, food and water sources, and trade routes, as opposed to attacking the enemy's soldiers ('regular warfare'). This 'irregular' warfare became the regular form of warfare for the United States when colonizing the continent and wiping out indigenous populations. This style of warfare is designed not for the winning of battles and defeating armies, but for the total destruction of the enemy's civilization. The goal of 'irregular' warfare is to win wars by any means necessary, through the elimination of other civilizations from the face of the planet if necessary, which is exactly what nuclear weapons do. Nuclear bombs would serve no better than regular bombs in attacking soldiers or battleships or airplanes, they are only useful for destroying population centres and killing civilians. It is exactly this type of warfare that the United States perfected as means for committing genocide against indigenous peoples, and it is exactly this type of warfare that the United States used to let loose the most dangerous weapon in the history of Japan in 1945.

Because nuclear weaponry is a tool for 'irregular' warfare, and the goal of 'irregular' warfare is the absolute destruction of your enemy's civilization, it should be the logical conclusion that a war between multiple nuclear powers where nuclear weapons were used would signal the end of civilization as a whole, and that was the fear from the end of Second World War until the fall of the Soviet Union in 1991. The world was dominated by two superpowers, the United States and the Soviet Union, with vast nuclear arsenals, and every time the two superpowers clashed there was a risk of nuclear war. This was prevented by the two superpowers exploiting the conflict in the global south to be used as proxy wars, such as the Korean War, the Vietnam War and countless others, in order to gain an advantage over the other in control of the world's people and resources. It was the Cold War and imperialism that fuelled the drive for uranium mining in the four corners region. It was the conflict between capitalism and socialism that created the incentive for capitalists to mine uranium, and it was the very techniques of warfare used for the genocide of indigenous peoples that created the weapons that their mined uranium would build.

Japan, as a country that has been a victim of nuclear weapons, and its three non-nuclear principles (non-production, non-possession and non-introduction of nuclear weapons) give it a unique role in the movement against nuclear weapons, yet it has consistently been pro-nuclear, both energy and weapons. Many of Japan's post–Second World War prime ministers have been pro-nuclear or had pro-nuclear cabinet members. This has led to Japan's inclusion in the United States 'nuclear umbrella'. The core of Japan's defence policy is nuclear weapons; the weapons themselves will be American, but their purpose is clear: the defence of Japan. Much like the United States, Japan's non-proliferation policy turns a blind eye to US-favoured countries such as India and Israel, while denouncing unfavoured countries such as Iran and North Korea (Abramsky 2010).

The United Kingdom, France, China and Russia are all engaging in nuclear weapons modernization programmes, meanwhile Israel maintains a nuclear force, and Pakistan and India are building up nuclear forces. The United States claims to be 'revising' its nuclear arsenal to be more effective in a 'Post-Cold War security environment' (ibid.). The fact that countries such as the United States are modernizing their arsenals serves as encouragement to those states without nuclear programmes or with relatively weak ones to develop their arsenals, especially those of which the United States may be unfriendly towards, such as Iran. Peer de Rijk, director of the World Information Service on Energy, said that

> everything [Iran] has done in past years is legal under any international treaty. Yet, the simple fact that it is not considered an ally of the Western world and its interests mean that the US and others have been considering a war against Iran. (Ibid.)

Among uranium-exporting countries, two of the largest exporters, Australia and Canada, have strict guidelines to ensure that the exported uranium is used for peaceful purposes only and not used for the development of weapons or in ways that could support the development of weapons. Over the past few decades, the concern has been that uranium intended for civilian fuel use will be used for the development of weapons; however, the opposite has come true: surplus weapons-grade uranium and plutonium have increasingly been used for fuel production rather than for weapons. Military uranium has become an increasingly large portion of the uranium used for commercial fuel productions, and will likely increase as the governments of the United States and former Soviet countries continue to follow through on disarmament treaties that have been being put in place since 1987 (World Nuclear Association n.d.).

NUCLEAR ENERGY

Nuclear energy is a type of energy that uses heat generated by splitting atom nuclei in order to generate steam to turn turbines, similar to fossil fuel energy. However, there are some key differences between fossil fuels and nuclear energy. The first is that fossil fuels generate greenhouse gases such as carbon dioxide when they are burned, causing a multitude of environmental and health problems such as global warming, ocean acidification and increased respiratory diseases. Nuclear energy releases little greenhouse gas throughout the process, and absolutely none during the actual production of energy. Fossil fuels are far less efficient than nuclear energy (though the issue of nuclear waste is still important).

Uranium is generally found in two types, U-238 and U-235, the latter of which makes up approximately 0.7 per cent of natural uranium. U-235 is the main type of uranium that is 'fissile'. The process of 'enrichment' separates the isotopes in order to concentrate certain isotopes. Most reactors require the amount of U-235 to increase from about 0.7 per cent to 3-5 per cent. By firing a neutron into the nucleus of a U-235 atom, it will split, producing more neutrons that hit the nuclei of surrounding uranium atoms and the process is repeated, producing vast quantities of heat energy. The heat energy then is used to turn water into steam, which turns turbines and generates electrical energy. This process is far more efficient than fossil fuels, as it uses far less fuel to produce more energy and less waste, and more reliable and consistent than other forms of alternative energy such as solar and wind.

In 1954, the AEC announced that nuclear energy would soon be 'too cheap to meter', opening the possibility of a future in which energy could be free for all to use. Nuclear energy would soon become a standard method of energy production and penetrate every household and industry in the country. By the late-1960s, every European country had a nuclear power programme, but also by that time the voices in opposition to nuclear power were also making themselves heard. Opposition to the development of nuclear energy, due to the risks of meltdowns and troubles with nuclear waste, continued for the next decade until, in 1979, the worst nuclear disaster in the US history occurred at the Three Mile Island Generating Station, an accident that would ultimately cost $1 billion in clean-up. The nuclear industry never recovered from the events of Three Mile Island, as 110 orders for nuclear power plants were cancelled after that, and the last order for a nuclear reactor actually built in the United States was placed in 1974.

At the present time, thirty-one countries have operating nuclear power stations; however only six of these are responsible for about 75 per cent of production: France, Germany, Japan, Russia, South Korea and the United States. A new reactor has been designed, called the 'European Pressurised Reactor' (EPR), to produce cost-effective nuclear energy at the two sites France and Finland, where they have been under construction, however they have proven to be incredibly costly and serious concerns over safety have been raised. Designers hope that the reactor will usher in a new 'nuclear renaissance' and make nuclear energy more competitive. It has been proposed by the United States that only a few selected countries be allowed to produce nuclear energy: China, France, Germany, India, Japan, the Netherlands, Russia, the United Kingdom and the United States. The list excludes all Islamic countries and most of the global south (and the only countries of the global south on the list are generally the US friendly). The countries excluded from the list would have to buy their nuclear energy from those who are privileged enough to produce it (Abramsky 2010).

Bernard Weinstein, an energy economist at Southern Methodist University in Dallas, Texas, said about nuclear reactors that China is pushing ahead big time with nuclear. They've got 26 gigawatts of nuclear plants under construction (Jung 2013). On top of that, ten plants are being built in Russia and six are being built in India. This recent increase in nuclear reactor construction has slightly increased demand for uranium, 'So if someone was looking to invest in mining uranium ore in the United States, it's probably going to be for markets abroad', said Weinstein (Jung 2013).

According to the OECD's 'World Energy Outlook', a significant increase in nuclear energy to replace fossil fuels would be required to maintain greenhouse gases below 450 ppm. Many environmental groups view 450 ppm as too high and hope to keep them below 350 ppm, which has already been surpassed and will require the removal of greenhouse gases from the atmosphere. One way to combat the release of greenhouse gases into the atmosphere is to replace fossil fuels with nuclear energy, because nuclear does not release any greenhouse gases during the energy production process (though mining and transportation and such do release gases). Throughout the whole process, including extraction, milling, enrichment, transportation and energy production, less greenhouse gases are released than in the life cycle of solar power production.

Despite its lack of greenhouse gas emissions, nuclear does present some very harmful and infamous problems to the environment. I've already talked about uranium mining and its effects on the environment and human health, but there are also problems with nuclear energy such as reactor meltdowns and leaks. Accidents such as the Three Mile Island disaster in 1979, the Chernobyl disaster in 1986 and the Fukushima disaster of 2011 are a few examples, and while nuclear disasters are not commonplace, the infrequency of their occurrence is perhaps outweighed by the severity of the disasters they cause. Nuclear meltdowns and reactor leaks cause serious environmental destruction and are harmful to human health in a capacity that is hardly seen anywhere else. In addition to this, nuclear waste has to be kept in isolation for anywhere from tens of thousands to hundreds of thousands of years. Ten thousand years ago we were just developing agriculture, can we imagine where we will be 10,000 years from now? The possibility of future societies not knowing the dangers of nuclear waste is why we have developed symbols for radioactivity and biohazards that we hope would be effective in warning future peoples of danger without having to use written language that they may not understand.

Currently, nuclear energy production, though rising, is not rising as fast as energy consumption, and as the climate worsens due to overuse of fossil fuels, the need for a form of energy that produces at the rate of fossil fuels is growing more and more. However, it is hardly possible to maintain the percentage of energy that nuclear energy makes up as energy demands increase, it is estimated that over the next ten years, eighty new reactors would have to be built simply to maintain its current share, let alone begin replacing fossil fuels. Further, as oil supplies decrease (though we are in no danger of running out before we experience ecological collapse), competition for such resources grows fiercer. And any model for increasing nuclear energy production has to take into account uranium supplies and current reactor capacity. It is estimated that sometime between now and 2030, uranium stockpiles will be depleted, and production cannot increase at a fast enough rate to make up for the increase in demand for energy. About 2.3 Mt of uranium have been produced and known remaining reserves are generally of lower quality and lower concentration so they may not be worth the energy it takes to extract them. If estimates of undiscovered uranium from the Nuclear Energy Agency are included, reserved double or quadruple, which gives some hope to the prospect of a uranium-fuelled future, but not much. The likelihood of these reserves being found is less than the likelihood of not being found, and they are far too speculative for any serious planning of what the future of energy is going to look like (Abramsky 2010).

The life cycle of a nuclear power plant is a long one, including several years of planning and at least five years of construction. The plant may then operate for several decades, with estimates being around forty years. Some 45 per cent of reactors have been operating for over twenty-five years, and about 90 per cent have been operating for more than fifteen. Current nuclear reactors should begin reaching the end of their lifetime around 2030, at which time they will need be replaced in order to maintain productive capacity, but serious planning for this time has yet to have taken place (ibid.).

Twelve countries – Argentina, Bulgaria, Congo, Czech Republic, France, Gabon, Germany, Hungary, Portugal, Romania, Spain and Tajikistan – have all depleted their uranium supplies, and the bulk of known uranium supplies remain in Kazakhstan, Canada and Australia. Only if we can construct adequate nuclear breeding reactors (which generate more fissile material than they consume) can we ensure that uranium supplies do not run out over the next twenty years. However, neither nuclear breeding reactors nor thorium reactors, which will be discussed further in this chapter, can be built in time to be cost-efficient and competitive in the market.

Other forms of alternative energy exist beyond nuclear that many readers will be aware of, and certainly they will play an increasing role in the future of our energy system, but as of now they all have problems that need to be solved before they can be put into use on the scale that we currently require. Solar power as a technology is consistently getting better but it has some serious drawbacks that get forgotten by people when imagining a green energy future. First, the mining of rare earth minerals for solar panels is especially harmful to the environment, as any mining is, and some minerals are scarce. The second reason is solar panels last only for a few years before they need to be replaced and are difficult to recycle. Finally, solar panels get incredibly hot, which can be seriously harmful to animals and plants near the solar farms. Another possible alternative is geothermal energy, which harnesses the earth's internal heat to use underground steam to turn turbines to generate electricity. Pockets of underground steam are rare though and can be created by injecting water into the ground, but this also has the possibility of creating many of the problems associated with hydraulic fracturing.

Even with these critiques, renewable energy future is not impossible; however, two things must be done in order to ensure that these forms of energy don't end up causing more damage. The first is that we must invest more time and energy into the research and development of renewable energy technology in order to produce it in ways that will be sustainable and environmentally friendly. The second is that we must make sure that while we are implementing these technologies, we must do it in ways that are safe for the communities they are providing for and extracting from, the people working to produce the energy and the planet as a whole. The capitalist drive to cut costs causes us to overlook safety and environmental impact in exchange for saving money, costs that can be externalized are thrust upon the most vulnerable communities. Centralized control over resources, present in hierarchical institutions, especially in capitalism, allows those with power to affect change to ignore the consequences of their actions while many suffer.

Since the birth of nuclear energy in the late 1940s and early 1950s, it has been proposed by scientists that we could produce nuclear energy with materials other than uranium, notably thorium. The idea to use thorium instead of uranium to produce energy was originally proposed in the late-1940s, but the government refused to fund research into it because the process burned excess plutonium, while the government needed the plutonium to build nuclear weapons during the Cold War. However, as the climate crisis grows more direr and alternative energy sources are not progressing fast enough to replace fossil fuels soon, we need to find a way to power our civilization that is both ecologically friendly and can meet our energy demands. Thorium-based nuclear energy solves these problems and more.

The first advantage of thorium is that it is abundant in nature. It is estimated that three times as much thorium on earth than uranium is available, and currently we have few uses for it, and it mostly ends up as a waste product, contaminating rivers after mining rare earth metals. The second advantage to thorium, especially considering how abundant it is, is that it produces far more energy than any other current energy source. One tonne of thorium can produce as much energy as 200 tonnes of uranium, or 3.5 million tonnes of coal. Also, all thorium is usable for energy, as opposed to the small percentage of uranium that is usable, and therefore does not require enrichment, which also saves time and energy in producing the fuel. Additionally, the waste produced by thorium-based nuclear power needs to be kept isolated only for a few hundred years, unlike waste products from uranium reactors which have to be kept isolated for up to hundreds of thousands of years. Thorium's melting point is also approximately 500°C higher than that of uranium, which provides an extra margin of safety against meltdowns in reactors. There are still the issues of the immense initial investment in the construction of power plants, and the issues of the impact of mining (though mining is likely to continue, so we might as well use the upturned thorium rather than let it contaminate rivers), but a dialogue around thorium and nuclear energy should be happening.

CONCLUSION

Capitalism has turned uranium from a nuisance into a valuable commodity. The capitalist drive for cheap sources of energy turned uranium into nuclear energy, but the need for profit kept capitalism from turning nuclear energy into a practically free, practically limitless resource for the people to use. Capitalist cost-cutting and externalizing damages have turned nuclear energy into one of the most taboo forms of energy in the world. The need to exploit impoverished peoples for cheap labour fuelled colonialism in the four corners region of the United States. Capitalism and production for profit have turned uranium, and all sources of energy, into questions of cost-efficiency and marketability rather than a question of resource-efficiency, safety and human need.

The 'irregular warfare' of American colonialism led to the creation of nuclear weapons out of uranium. The possibility of nuclear immolation is all-too-real of a fear, and it is one that is the culmination of over 500 years of colonialism. And not only did colonialism fuel the conception of the weaponry, but it is through colonialism that people were able to obtain the uranium to build the weaponry. Exploiting the labour of Navajo people in the four corners and of African people in Shinkolobwe would fuel the uranium bombs that would be dropped on Japan in Hiroshima and Nagasaki, as well as the nuclear arsenals of the superpowers that played a dangerous game of chess with 'third world' countries during the Cold War.

Uranium has the potential for so much good and so much harm to the biosphere. The radioactivity and chemical toxicity of uranium requires that it be carefully handled and used so that it does not cause damage to ecological and human health. But currently the prospects of replacing fossil fuels otherwise are looking slim. Institutions such as capitalism and the state have made nuclear power dangerous and destructive, by producing power for profit rather than for safety, to meet human needs and to liberate us from the chains of fossil fuels. If we wish to develop other forms of alternative energy to their fullest capacity, and if we wish to use nuclear energy to safely power our civilization, we must abandon centralization and the profit motive, and we must produce communally, through participatory planning, and for human need.

Uranium has a long, complex and dark past: a past of colonialism, imperialism and environmental destruction. But uranium is not inherently evil, it is merely a radioactive element with ninety-two protons and isotopes ranging from U-233 to U-238. It is through the process of commodification and exploitation for the benefit of the few at the expense of the many that uranium, like many other things, becomes a tool for evil. Uranium may have a dark past, but if we can move past capitalism and promote ideas of non-hierarchical social organization and ecological principles, perhaps it can have a better future (or perhaps it will be abandoned altogether, but a dialogue must happen, and we must think outside of capitalism while it happens).

NOTES

1. Capitalism further relies on the denial of vital resources to the masses in order to sell them and make profit.

REFERENCES

Abramsky, K. (2010). *Sparking a Worldwide Energy Revolution: Social Struggles in the Transition to a Post-Petrol World.* AK Press.

Amundson, M. A. (2004). *Yellowcake Towns: Uranium Mining Communities in the American West.* University Press of Colorado.

Brugge, D., Benally, T., & Yazzie-Lewis, E. (2007). *The Navajo People and Uranium Mining.* UNM Press.

Dahlkamp, F. J. (2013). *Uranium Ore Deposits.* Springer Science & Business Media.

Dunbar-Ortiz, R. (2014). *An Indigenous Peoples' History of the United States.* Beacon Press.

Evans-Pritchard, A. (2010, August 29). *Obama could kill fossil fuels overnight with a nuclear dash for thorium.* http://www.telegraph.co.uk/finance/comment/7970619/Obama-could-kill-fossil-fuels-overnight-with-a-nuclear-dash-for-thorium.html.

Ferguson, C. D. (2011). *Nuclear Energy: What Everyone Needs to Know.* Oxford University Press.

Frosch, D. (2009, July 26). Uranium contamination haunts Navajo Country. *The New York Times.* http://www.nytimes.com/2009/07/27/us/27navajo.html.

Hammond, C. R. (2000). The elements. In *Handbook of Chemistry and Physics* (81st ed.). CRC Press. https://www-d0.fnal.gov/hardware/cal/lvps_info/engineering/ elements.pdf.

Johnston, B. R. (2007). *Half-Lives and Half-Truths: Confronting the Radioactive Legacies of the Cold War.* School for Advanced Research Press.

Jung, C. (2013, December 21). *Navajo Nation opposes uranium mine on sacred site in New Mexico.* Retrieved November 29, 2016, from http://america.aljazeera.com/articles/2013/12/21/navajo-nation-opposesuraniummineonsacredsiteinnewmexico.html.

Kazimi, M. (2003). Thorium fuel for nuclear energy. *American Scientist, 91*(5), 408. https://doi.org/10.1511/2003.5.408.

MacDonnell, L. J., & Bates, S. F. (2010). *The Evolution of Natural Resources Law and Policy.* American Bar Association.

McHarg, A. (2010). *Property and the Law in Energy and Natural Resources.* Oxford University Press.

McLeod, C. (1983). *The Four Corners: A National Sacrifice Area?* [VHS].

Merkel, B. J., & Hasche-Berger, A. (2008). *Uranium, Mining and Hydrogeology.* Springer Science & Business Media.

Merkel, B. J., & Schipek, M. (2011). *The New Uranium Mining Boom: Challenge and Lessons Learned.* Springer Science & Business Media.

Moore, J. W. (2015). *Capitalism in the Web of Life: Ecology and the Accumulation of Capital.* Verso Books.

Rodney, W. (1978). *How Europe Underdeveloped Africa.* Bogle-L'Ouverture Publications.

Roscoe, R. J., Deddens, J. A., Salvan, A., & Schnorr, T. M. (1995). Mortality among Navajo uranium miners. *American Journal of Public Health, 85*(4), 535–540.

Spitz, J. (2009). *The Return of Navajo Boy* [DVD].

Voyles, T. B. (2015). *Wastelanding: Legacies of Uranium Mining in Navajo Country.* University of Minnesota Press.

World Nuclear Association – World Nuclear Association. (n.d.). Retrieved November 29, 2016, from http://world-nuclear.org/.

Zoellner, T. (2009). *Uranium: War, Energy, and the Rock That Shaped the World.* Penguin.

3

Moving beyond Borders

Anarchist Political Ecology and Environmental Displacement

Nicolas Parent

Of the many forms of violence people across our planet endure, the physical production of borders is one of its most pernicious and unappreciated iterations. They have been used as a political exercise in defining the 'we', and rallying the will of independently minded and autonomous individuals to a paradigm committed to the exclusion of the 'other'. While political borders have defined the geographic delimitation of nations, legitimizing the state as we know it, the borders separating space into private and public have legitimized the spatial dominance desired by people, institutions and corporations who seek to exert power over others. As such, the human and natural conditions have come to be defined, in part, by the 'in' and 'out' of the borders of everyday life.

Forced migrants, in the form of asylum seekers, refugees and internally displaced persons (IDP), are one of the several groups who experience the impacts of the bordering process. While the actions of the state and their neoliberal partners cooperate in the production and ripening of conditions which cause displacement, the human tragedy of forced displacement is met with attitudes and policies that dehumanize and criminalize those who have been displaced and dispossessed. Environmental migrants, yet another iteration of forced displacement, are a manifestation of such circumstances; the result of a world that has been increasingly privatized, bordered and toxified. As the following chapter will argue, the latter holds strong argumentative purchase for the deconstruction – and complete elimination – of the international border regime. It will do so through an approach anchored in political ecology, and one specifically leaning on its anarchist tradition.

As political ecology has seldom been applied to the context of forced migration, this chapter follows a logic which first and foremost builds the practical and theoretical foundations necessary to understand its applicability to the situation of environmental displacement. The first section outlines the current state of solidarity with forced migrants at a global scale, followed by the second section which specifically presents the current debate on environmental displacement. The third section introduces political ecology as a novel and important approach to understanding the conditions which produce environmental migrants relying on anarchist roots to engage in a discussion on power and domination. Finally, the last section applies the normative function of political ecology, leveraging ideas from anarchists such as Élisée Reclus and Peter Kropotkin to call for the absolute deconstruction of borders.

THE PRECARIOUS STATE OF SOLIDARITY WITH FORCED MIGRANTS

At the dawn of the twentieth century, the French geographer and anarchist Élisée Reclus described the state of humanitarianism as an object of ridicule by statesmen and writers for its 'poor sentimentality'. While Reclus recalled the revolutionary ideals about progress developed during the second half of the nineteenth century, he concluded that 'these brave souls had no idea of the difficulties that their propaganda would have to encounter' (Reclus 2004). The idea of 'humanity' was, in fact, falling victim to the isolationism of nation-state, 'A passionate nationalism rages in all western countries and existing borders have for the most part been tightened during the past fifty years' (ibid.). Unbeknownst to Reclus, this nationalism, of which only gained momentum following his death, would result in wars of unprecedented scale and some of the most severe tragedies in contemporary human history.

Following the mass migration of Belgians, Serbians, Russians and Armenians resulting from the First World War and over a million refugees as a result of the 1921 Russian famine, the very first refugee-centred multilateral organization – High Commissioner for Refugees (HCR) – was formed. By the end of the Second World War, a staggering 1.2 million people had been displaced (Hansen 2014). With the gravity of these facts, where 'humanitarianism and refugees entered into an increasingly co-dependent relationship' (Barnett 2014), the United Nations incorporated HCR within its organizational umbrella, forming the United Nations High Commissioner for Refugees in 1950. In 1951, the Convention for the Protection of Refugees became the very first international legal instrument designed to assert the protection and rights of those fleeing persecution. After several other refugee-producing conflicts erupted across the globe, it became clear that the Refugee Convention needed to be amended in order to make it suitable internationally. Thus, in 1967 came the Protocol Relating to the Status of Refugees, building onto its predecessor. At this time in history, refugees were still welcomed with good will in order to exhibit generosity and humanitarian values in the West, on a backdrop of the Cold War.

After the fall of the Berlin Wall, effectively marking the end of the Cold War, attitudes towards refugees began to shift. As Loescher (2014) notes,

> The period since the Cold War has been marked by a shift from asylum to containment where Western states have largely limited the asylum they offer to refugees and have focused on efforts to contain refugees in their region of origin. (Loescher 2014)

In contrast to a time when 'they were once content to react on the basis of obligation and expectation' (Goodwill-Gill 2001), states 'introduced a wide range of measures designed to keep asylum seekers from reaching national borders' (Hansen 2014, 258) or, at best, within their respective territories or regions.

The incremental distance that has occurred between states in the 'developed' world and their responsibility towards those fleeing global conflicts is understood as the emergence of what Akram (2015, 2016) calls the 'containment paradigm', and what is believed here to be the feedback effect of the growth of risk analysis and risk aversion within industrialized nations.[1] Effectively, the result has been what is commonly understood as the 'securitization of migration' (Lazaridis and Wadia 2015; Humphrey 2014; Gabrielli 2014) or rather the trend of perceiving and representing migrants as threats to national security, employment structures and social cohesion, among others. The wide use of refugee-related 'water metaphors' in political, public and media discourse – engendering feelings of crisis, danger and refugees as problems – exemplifies one of many ways that these perceptions have been displayed (Parker 2015; Pickering 2001; Baker and McEnery 2005; Kainz 2016). To properly enunciate the mechanics of the containment paradigm and its physical and ideological endorsement by the Global North, the 'European refugee crisis' case study is outlined in the chapter, starting from the process of externalization.

Externalization is understood as 'the process which uses various methods to transfer migration management beyond national borders [and] in the EU involves transferring responsibility and, in effect outsourcing its immigration and asylum policy by subcontracting controls' (Rodier 2013). The use of externalization is nothing new for the European Union with the use of visas, carrier sanctions and airport international zones attesting to this fact (Hansen 2014). Readmission agreements, yet another and arguably a more cunning form of externalization, show a more complex interaction between diplomacy, foreign relations and development. In principle, readmission agreements essentially outline the 'procedures for one State to return aliens in an irregular situation to their home State or a State through which they passed en route to the State which seeks to return them' (Muedin 2010). With a basic understanding of migration flows, whereby the top refugee-producing countries are located in the less developed Global South, these agreements tend to favour and be called upon by developed countries in the Global North. This asymmetric agency between 'developed' and 'developing' parts of the world is further evident in the way that readmission agreements are often part of larger development agreements, questioning the authenticity and intentions of international cooperation (Migreurop 2012). These agreements have proven to be effective in returning 'irregular' migrants to their country of citizenship, irrespective of their precarious sociopolitical climate.[2]

This has been the case for the deportation of irregular migrants back to Pakistan by means of the 2010 European Union–Pakistan readmission agreement (Carrera 2016), the return of Syrians to Turkey through the 2016 European Union–Turkey Agreement (Parent 2016), and of deportation of Afghans from the European Union through the 2016 Joint Way Forward (JWF) agreement (Rimmer 2017). One final notable form of externalization is that of 'pushbacks': the forcible return of migrants into a territory they are fleeing, albeit in direct violation of the principle of *non-refoulement*. Although extensive reports by Amnesty International (2013) and Pro Asyl (2013) have shed some light on the wide extent of pushbacks in the Aegean Sea by the Greek coast guard, other reports indicate that this practice is widely used at Europe's external borders in the Mediterranean[3] and Balkans.[4]

Another means that securitization and containment paradigms have manifested themselves is through the growth of the militarization of migration management. This is most evident in the mass increase of funding of Frontex – the European Union border management agency – where it has gone from receiving an annual median of 90 million euros between 2009 and 2014 to over 300 million euros in 2017. These funds have predominantly been allocated to Frontex' maritime joint operations in the Mediterranean (Operation Triton) and Aegean (Operation Poseidon) Seas, tasked with intercepting boats of irregular migrants.[5] Another programme that has seen an increase in financial allocations has been the European border surveillance system – commonly known by the acronym EUROSUR. Complimentary to military surveillance operations, an increasing focus on EUROSUR's sophisticated early detection technologies marks a vivid trend towards the technologization of migration management (Martin 2011). Furthermore, and similar to the efforts by the Australian government to put in place its 'Pacific Solution' (McKay 2013), the military-industrial complex of the European Union's approach to migration has resulted in the expansion of its detention facilities for irregular migrants, or by funding the construction of military-grade holding facilities in EU candidate countries such as Serbia and Macedonia.[6] As Bosworth (2008) suggests, these areas potentially serve more than so-called 'administrative purposes':

> Prisons or immigration removal centres are singularly useful in the management of non-citizens because they enable society not only physically to exclude this population, but also, symbolically to mark these figures out as threatening and dangerous. (Bosworth 2008, 207–208)

Brought under mass condemnation by a wide array of international agencies, (Human Rights Watch 2017; MSF 2014; Leghtas 2017), these prison-like facilities have doubled as military outposts for a region that is increasingly known as 'Fortress Europe' (Carr 2016).

Only a small window into the containment paradigm, Europe's response to migration exemplifies the degree to which the Global North has lost its sense of 'humanity', to quote Reclus. Aside from having detrimental consequences on those fleeing crisis and conflict externally, the ramifications of Europe's anti-immigration discourse has led to a continental implosion, where some of its fundamentally collectivist regional arrangements, such as free travel and work permits across the Schengen zone, an embrace for multiculturalism and ease of trade between European countries are quickly eroding, as has been seen with the reinstating of borders in Germany and Austria, the rise of nationalist movements in Greece and Hungary, and the United Kingdom's decision to leave the European Union.

With similar policies and discourse in other developed nations such as the United States (Young 2017) and Australia (UNHCR 2017), the containment paradigm exhibited by the European Union is seemingly the rule, rather than the exception. With this in mind, it is no wonder that 'the average duration of a refugee situation has nearly doubled in the past decade to a staggering 18 years' (Loescher 2014, 225). All of these facts are worrying when considering that the number of forcefully displaced people has been steadily increasing. According to UNHCR data for the last ten years (2007–2016), the number of asylum seekers has nearly quadrupled, refugees have seen a 50 per cent increase, and there are 2.7 times more IDPs.[7] Furthermore, and of particular relevance to this chapter, is that international organizations such as the International Federation of Red Cross and Red Crescent Societies (IFRC), UNHCR and the Internal Displacement Monitoring Centre have all acknowledged that displacement due to climate-related disasters is on the rise (International Federation of Red Cross and Red Crescent Societies 2009; Yonetani 2015; Padoan 2015). However, unlike refugees and IDP's, environmentally displaced persons have no formal recognition within the international community and considering the current political landscape that exhibits hostility towards forced migrants, it seems unlikely that they will in the near future, or that their recognition would even lead to a fruitful, sustainable and dignified end.

CURRENT DEBATES ON ENVIRONMENTAL DISPLACEMENT

Taking centre stage in the conversation on environment and displacement is the fierce debate surrounding the term 'environmental refugee'. Although the term has been used as early as the 1970s, the term 'gained substantial popularity' and recognition following El-Hinnawi's 1985 report *Environmental Refugees* for the United Nations Environment Program (UNEP). Here, El-Hinnawi provided a definition of the term:

> Environmental refugees are those people who have been forced to leave their traditional habitat, temporarily or permanently, because of a marked environmental disruption (natural and/or triggered by people) that jeopardized their existence and/or seriously affected the quality of their life. By 'environmental disruptions' in this definition is meant any physical, chemical and/or biological changes in the ecosystem (or the resource base) that render it, temporarily or permanently, unsuitable to support human life. (Cardy 1994)

Combined with a political backdrop that was increasingly engaged in the discourse on global environmental degradation, El-Hinnawi's UNEP publication became the catalyst to a growing number of literary works discussing the topic of 'environmental refugees'. Most of these accounts were based on neo-Malthusian principles, attributing displacement to resource scarcity as a result of population growth.[8] For instance, in making an attempt at explaining root causes of environmental displacement due to soil degradation, Otunnu (1992) attributed the mismanagement of land in sub-Saharan Africa to an inadequate and overly saturated education system, using this as leverage to urge national government to devise '(. . .) coherent and informed policies to control high population growth' (Otunnu 1992).

These initial perspectives were the early foundations of what is widely understood as the 'maximalist' position on the question of environmental refugees. Studies produced by academics such as Jacobson, estimated the number of environmentally displaced persons at 10 million (1998), and Myers and Kent who raised this number at 25 million in 1995 (1995), and predicted over 200 million environmental refugees by 2050 due to rising sea levels and drought (Myers 2002). These theories gained substantial political traction in the technocratic bureaucracies of the Global North. Effectively, Myers' studies and 'his conception of the "environmental refugee" has become fundamentally important in shaping how the term has been taken up amongst academics, policymakers and the public' (Morrissey 2012, 37). As a result, the maximalist's vividly alarmist voice has succeeded in advancing two notable political agendas. On the one hand, the fear of a growing number of forced migrants due to environmental stress has been utilized to advocate for more proactive environmentalist policies in the Global North (ibid, 41). On the other, this same fear has been used to frame climate change and migration within a wider discourse on global security,[9] further supporting the aforementioned securitization paradigm by considering climate change as a 'threat multiplier' (Trombetta 2014).

In contrast to the maximalist perspective, Suhrke (1994) describes the 'minimalist' view as one that sees environmental change as a 'contextual variable that can contribute to migration, but analytical difficulties and empirical shortcomings make it hazardous to draw firm conclusions' (474). Effectively, minimalists believe that environmental pressures can lead to an exacerbation of, or intersections with, current instabilities that may cause forced migration, such as armed conflict (Kibreab 1997; Castles 2002). This has led some, such as Black et al. (2011), to devise frameworks outlining the different drivers of migration, allocating a special emphasis on the influence of environmental change on political, demographic, economic and social drivers at the macro scale (Black et al. 2011). Yet, as opposed to the majority of maximalist/alarmist scholars of whom predominantly engage with disaster research, ecology and conflict studies, the migration scholars of the minimalist/sceptic position (Dun and Gemenne 2008) have been wary of accepting the term 'environmental refugee' as one that can or should be granted international protection (Kibreab 1997, 21). This is mainly predicated on a legal argument that, effectively, current protection instruments do not consider environmental stress as a legitimate cause for forced migration (Boano et al. 2008). Yet, this produces an undeniable paradox considering that migration scholars (Long 2013), not to mention minimalists (Zetter and Morrissey 2014; Suhrke 1994), have themselves underlined the multi-causal nature of forced migration albeit its legal recognition grounded largely in a mono-causal process of refugee status determination.

Currently, both 'maximalist' and 'minimalist' perspectives stand fixed in a grid-lock, where maximalist estimates, such as Myers', have been widely undermined, albeit for some good reasons, and where minimalists, such as Black (2001), have confirmed that although most agree that environmental change can sometimes be the cause for involuntary migration, 'practical concern with the plight of poor people leaving fragile environments has not translated into hard evidence of the extent or fundamental causes of their problems' (Black 2001), making a concrete definition of 'environmental refugee' challenging and thus policy nearly impossible. Here enters the need for a fresh approach that can resuscitate the conversation, providing a new analytical approach to confront root causes for both environmental change and migration at a range of scales, and new avenues to facilitate human movement by deconstructing the institutions that have legitimized anthropogenic environmental degradation and upheld inherently antisocial barriers to livelihood, dignity and means of responding to environmental change. In the following sections, it will be shown that political ecology, and more specifically one anchored in an anarchist foundation, can act as an effective springboard to reframe the variables at play in the debate on environmental displacement, and ultimately move towards actionable solutions with human dignity and environmental integrity in full consideration.

TOWARDS AN ANARCHIST POLITICAL ECOLOGY OF ENVIRONMENTAL DISPLACEMENT

In the final sentence of their introduction to the 1994 inaugural issue of the *Journal of Political Ecology*, Greenberg and Park attest to the following: 'we feel it would be ill-advised to define "political ecology" and maintain rather that all legitimate forms of political ecology will have some family resemblances but need not share a common core' (Greenberg and Park 1997, 8). With relative agreement, but attempting to give some level of structure, Robbins (2012) provides a general and open-ended characterization of political ecology as a community that assembles around certain kinds of texts which seek to 'address the condition and change of social/environmental systems, with explicit consideration of relations of power' (20). As a result, political ecology is effectively based on an interdisciplinary and critical approach that is nestled between social and natural sciences, attracting a vast diversity of intellectuals, researchers and practitioners from a mosaic of disciplines. In the case of a political ecology anchored in an anarchist foundation, it considers that the aforementioned relations of power are, effectively, the result of a now globalized predatory system whereby the domination of both people and nature are central to the fundamental existence of the stated system. As these forms of domination are central and indivisible to the current 'world order', anarchist political ecologists do not seek incremental (i.e., reformist) changes in the current system, but rather, normatively call for the *absolute dismantling* of systems and institutions that exert power upon both – independently and interactively – human and natural communities.

As mentioned previously, the discussion on environmental displacement is effectively at a standstill, with one major impediment to its progress being, as Morrissey (2012) asserts, that the conversation has been more about '*how a relationship is represented*, and less about the *nature of the representation itself* (Morrissey 2012, 40, original emphasis). Effectively, the evidence presented in the literature on the interaction between environment and migration has been predominantly used to advocate for the use or misuse of 'environmental refugee' as a term or status recognized by individual states and the international community. On the one hand, some have argued against expanding the definition of 'refugee', claiming that the links between environmental stress and migration are tenuous (Black 2001), or based upon the fear that amending current legal instruments would undermine rights currently afforded to those already recognized by hosting states (Keane 2004). On the other hand, authors such as Conisbee and Simms (2003) have advocated that the environment be included as a cause of persecution under international law, predicated on the assertion that we now have sufficient scientific understanding of the role played by human activities in the development or perpetuation of environmental issues such as climate change. Specifically, they argue that government policies or development projects can use the environment as an 'instrument of harm' (Conisbee and Simms 2003, 30), knowingly putting people's livelihoods in danger. Effectively, this perspective closely tracks with what Beck (1992) has called the 'reflexive modernity' where there is an increasing demand put onto risk-distributing political apparatus' to respond to conflicts, crises and disasters to which they themselves are responsible in producing. Yet, in trying not to discredit the well-founded observation that environmental persecution is both real and an extensive driver of displacement, status-based solutions, such as those by Conisbee and Simms (2003), exhibit a naïve understanding of the political landscape in the thematic milieus of both environment and migration. As exemplified in one of the most irrefutable cases of environmental-forced migration – the movement of islanders in the Pacific as a result of climate-related sea level rise – McNamara (2008) demonstrates that by crafting bilateral resettlement agreements directly with island states rather than engaging in multilateral discussions on the impact of climate on migration, the Global North has been largely successful in averting an internationally recognized culpability for the transnational impacts of its environmental pollution. Effectively, to recognize environmental displacement, not to mention the inclusion of the environment as a cause for refugee status under international law, is unlikely in a world where state and private interests spreading destructive neoliberal agendas are increasingly bound and defined by a general ethos of blame-avoidance.[10]

As demonstrated, Morrissey's (2012) diagnosis of the debate on environmental refugees is indeed useful, as it underscores the myopic analytical imagination of those who have sought to comment on the interaction between environmental stress and forced migration. Rather than focusing on the nature of this interaction and how broader systems of exploitation, injustice, marginalization and violence participate in the reproduction of environmental-forced displacement, the vast majority of authors have become bogged down in a debate on nomenclature and legal jargon. Persons who have been displaced due to environmental stress or degradation are, indeed, living testaments of an ecology that has been privatized, bordered and toxified, and therefore their cases need to be treated as such. As the 'analysis of *power* remains central in political ecology' (Walker 2006, 388 emphasis in original text), it should be understood that asymmetries in power are manifested and reproduced through systems of private property, the process of bordering and environmental racism that is licensed by both the state and capitalist enterprise. Thus, the case of environmentally induced forced displacement is a particularly ripe and fruitful subject of analysis from the perspective of anarchist political ecology. That said, Robbins (2012) identifies three notable ways in which political ecology is different from traditional – or *apolitical* – forms of problem identification and resolution. These will be used to structure the remainder of this chapter.

Political ecology seeks to problematize specific social/environmental issues by 'identifying broader systems rather than blaming proximate and local forces' (Robbins 2012, 20). This ethos works particularly well in the case of environmental-forced displacement considering that inherently antisocial broader systems are tightly intertwined with those that are intrinsically anti-environment. The systemic rationalizations that recklessly perpetuate cycles of poverty, fracture solidarities and push communities to marginal geographies are hardly any different than those that enable state and industry to operate without hindsight or foresight of ecological damage, embodying anthropocentric conceptions of ecosystems as providing 'environmental services' to humankind and violently reorganizing landscapes based on function and interest. For this reason, the simple claim (and others like it) that the 'lack of dialogue between ecologists and social scientists render the links between environmental change and forced migration complex and debatable' (Boano et al. 2008, 4–5) seems nothing short of absurd. As previously stated, environmental refugees are one of the many visible manifestations of a world that has been increasingly privatized, bordered and toxified. In the case of privatization, neoliberalism's deep embedment in both private and public institutions is undoubtedly the driving force of its momentum, and the extent of its influence should not be undermined:

> The emergence of neoliberal ideology and its consolidation as the dominant economic system has radically reshaped the globe, intensifying already existing uneven geographies and resulting in a new level of complexity as established political structures, modes of governmentality, identity categories, economic matrixes, subjectivities, institutional frameworks, juridical processes, and epistemological positions are all being remade [. . .] creating new and unforeseen constellations of exploitation and struggle. (Spring and Ince 2012, 1592)

Neoliberalism, thus, is not exclusively an economic project, but one that is so invasive in its very nature that it transcends to the cultural, social, political and environmental spheres of everyday life. For instance, in the case of neoliberalism's 'triumph'[11] as the standard mechanism of 'development' in Africa, the output of its activities has done just the opposite, pushing millions into new forms of slavery through the processes of 'dispossession, displacement, and destitution' (Lebaron and Ayers 2013), while simultaneously an anti-corruption crusade has been launched by the Global North. This mission is considered a legitimate response to the continued political phenomenon of concentrated power and wealth across the continent, the lack of multiscalar analyses – intentional or not – has meant that there has been a notable 'lack of recognition [. . .] of the opportunities for corruption that privatization and liberalization have embodied over the past two decades' (Brown and Cloke 2004, 291). In respect to the question of environmental displacement, the effects are clear in global examples such as West Africa (Hammer 2004), the Greater Mekong (Sims 2015) and the Amazon Basin (Renfrew 2011), where state and internationally sanctioned neoliberal projects have unabashedly pushed subaltern communities to live in increasingly devoid, depleted and damaged environments. Simultaneously, the bordering process has thrived, making it increasingly difficult for these communities to find refuge. This has been the case for rural Cambodians who have increasingly lost communal lands as a result of widespread land grabs driven (in part) by foreign direct investment (Scurrah and Hirsch 2015; Davis 2013). Increasingly using migration to cities and abroad as a means to adapt to these forms of dispossession, they have been met with vividly anti-immigrant policies in neighbouring countries (Bylander and Reid 2017). Haselsberger (2014) explains that 'borders either confirm differences or disrupt units that belong together by defining, classifying, communicating and controlling geopolitical, sociocultural, economic and biophysical aspects, processes and power relations' (Haselsberger 2014), adding that contrary to our assumption that borders necessarily require a physical barrier demarking *in* and *out*, visible and invisible borders that demarcate people and places are found along a spectrum of 'thin' and 'thick' boundaries. Consistent with political ecology's commitment to a multiscalar approach (Neumann 2009), the bordering process has very direct and manifested impacts in the case of environmental displacement. Political borders as we know them do have an origin per se, being largely the result of the European unification process that was then exported – both physically and spiritually – to the rest of the world through colonial dominance[12] and nationalist independence movements.[13] Yet, and as evidence provided in the first section of this chapter suggests, both systems of borders and asylum have evolved to be equally vibrant, violent and codependent symbols of state dominance and control. Just as the process of bordering, control and exclusion through asylum has been adopted by the Global South:

> states within the developing world, many of them victims of the anti-migration regimes of the global north, have come to replicate the North's policies and popular rhetoric at a regional level [whereby] much more restrictive migration policies and practices have been adopted in many countries in the region. (Tobias 2012, 2)

Thus, where the Global South has indeed become the prey of the Global North's neoliberal interests in all its destructive forms, it has simultaneously imposed political barriers at a regional level, preventing people from escaping these aggressions of economic, cultural, social and environmental consequence. Communities have therefore been restricted to smaller scales of mobility which, as has been shown previously, have themselves been internally bordered through the incremental enclosure and private acquisition of public/community land. Effectively, as Walsh (2014) notes in his reflective piece on the political ecology of borders, 'even if we don't cross national borders every day, we are constantly moving along and across borders of private property. This scale of bordering space [. . .] often escapes scrutiny' (Walsh 2014, n.p.). Borderlands should, therefore, be conceived as much more nuanced than the spaces that separate states. In one notable way, the process of 'modernization' has forcibly advanced a model where space is bifurcated into rural and urban, pejoratively putting traditionalism and 'backwardness' against the fallacy of urban riches and opportunities. IDPs, recognized by some as largely resulting from the forces of development and climate change (Kälin 2014), have become increasingly intertwined with the process of urbanization, further exacerbating their invisibility as they become consumed by widespread urban poverty (Davies and Jacobsen 2009). As it should be recognized that the current literature has made some important contributions on the direct impact of slow-onset and sudden environmental changes as push factors of mobility, this last section has provided some additional signposts of broader systems contributing to the production of environmental-forced migrants.

Robbins (2012) attributes to political ecology the practice of considering 'ecological systems as power-laden rather than politically inert' (20). This is in direct response to other fields that have largely overlooked the 'political sources, conditions and ramifications of environmental change' (Bryant 1992, 13). Within the context of environmental displacement, Bristow explains that

> political ecology operates with the assumption that environmental problems are fundamentally related to social problems that stem from how societies are organised, who determines access to resources and how environmental costs and benefits are distributed. The state in this view does not only react to the problems associated with environmental change, but because it is deeply integrated with contemporary social, economic and political systems the state is implicated in the production of environmental problems. In this way, environmental displacement is, at its core, a political problem because it is these socially-determined factors, manifested in environmental degradation, which drive people from their homes. (Bristow 2007, 14–15)

Therefore, environmental refugees are effectively one notable ramification of asymmetric conditions of power created by political processes largely influenced (if not controlled) by agents which gain from – directly or not – the exploitation, acquisition, privatization or destruction of natural environments. It is no longer arguable that the natural environment is considered apolitical when taking into account the fact that ecological communities have been blatantly objectified rather than treated as an assemblage of intricate networks of organisms and species. As a result, it is those who have maintained harmonious and mutual relationships with these biological communities for hundreds (if not thousands) of years – or 'ecosystem people' (Gadgil and Guha 1995) – that have faced the brunt of resource depletion, habitat loss, toxification and, finally, displacement, as a result of the historical and contemporary processes that have redefined humanity's relationship with nature as one that is anthropocentrically parasitic.

For the last of Robbins' (2012) interpretation of political ecology's distinct approach – that of 'taking an explicitly normative approach rather than one that claims the objectivity of disinterest' (20) – the next and final section of this chapter will situate environmental displacement within a traditionally anarchist and normative post-border paradigm.

AN ANARCHIST CALL FOR MUTUAL AID
AND BORDER TRANSGRESSIONS

It has been the goal of this chapter to show how two significant issues have come to a head: first, the widespread moral precarity vis-à-vis our common responsibility towards forcefully displaced peoples and, second, how the greed and power of a few have malignly engaged with our common home – Earth – to exploit, deplete, destroy and, most importantly here, displace. Political ecology has proven to be a strong perspective to analyse the phenomena of environmental displacement, both as a result of its broad perspective and given that this particularly vibrant display of the environment as a political (rather than apolitical) space. Political ecology and anarchism make for a particularly rich partnership as they both have a historical gravitational interest around the principles of power and scale, along with a normative desire to 'unpack' these principles as a means to redefine, redress and repair them for the benefit of the Earth and all of its inhabitants. It was largely the friendship of two fascinating minds, both spatially minded, environmentalist and anarchist, that built the foundational blocks of political ecology (although it was not known as such at the time). These two friends and collaborators[14] were Élisée Reclus and Peter Kropotkin.

An outspoken anarchist and one that had a deep appreciation for nature, the work of Élisée Reclus was unique and controversial in an age of colonialism, industrialization and urbanization. The infamous image of hands holding up a globe, with it, inscribed the maxim *L'homme est la nature, prenant conscience d'elle meme* (Humanity is nature becoming self-conscious) was boldly selected for the preface of the first volume of *L'Homme et la Terre* (1905), in which Reclus wrote,

> It is the observation of the Earth that explains the events of history, and one that brings us back to a deeper study of the planet, towards a more conscious solidarity of our individual self, both small and large, with the immense universe.

As such, Reclus strongly believed that the relationship between humankind and nature needed to be mended, first and foremost, by a conscious awakening that destruction and dominance within humankind transcended to the natural world. As Reclus wrote in an 1866 publication, 'Among the causes in human history which have already continued to the disappearance of many successive civilizations, one must mention the brutal violence with which the majority of nations have treated the nourishing earth' (Reclus 1995, 33). Thus, his vision for a completely liberated society from all forms of dominance and hierarchy (Reclus 1894) was also one that inspired a vision for a natural world liberated from the dominance, abuses and destruction caused by humankind.

This was closely linked to Peter Kropotkin's work, which was conscious of the fragile equilibrium of natural systems and the need for humankind to live in harmony with its natural surroundings.[15] Similar to Reclus, who 'challenged the notion that contemporary social structure and ecological practice were the inevitable products of evolutionary selection' (Robbins 2012, 30), Kropotkin considered society as part of nature and the wider animal kingdom, and thus part of a 'social organization that is characterized more by mutual aid and reciprocity than by Darwinian struggle and competition' (Hall 2011, 378). Using the principle of mutual aid, which demonstrated that 'in the long run the practice of solidarity proves much more advantageous to the species than the development of individuals endowed with predatory inclinations' (Kropotkin 1902/1955, 17–18), Kropotkin sought to develop a type of eco-anarchism whereby the concept of mutual aid as part of nature and humankind 'would not impose sanctions or obligations on anarchists but it would put them in touch with the natural anarchist tendencies of the masses and would teach them to tailor their actions to its movements' (Kinna 1995, 282).

Linking the work of both Reclus and Kropotkin to forced displacement as a result of environmental changes helps to illuminate the relations of power and predation within human societies and the perversion of nature as an object of conquest. Both Reclus and Kropotkin opposed this form of domination, which has constructed the divisions between communities, dehumanized those that exercise freedom of movement, and brought about the environmental changes that force people out of their habitual and historical habitat. Environmental-forced migrants are indeed a manifestation of a humanity that is increasingly driving itself into a range of catastrophic outcomes, both for the survival of itself and others within nature. As Reclus animatedly noted that

> barring great cosmic revolutions whose shadows have yet to fall over us, modern nations will in the future escape the phenomena of seemingly final ruin that occurred to so many ancient peoples. Certainly, political 'transgressions', analogous to marine transgressions on coastlines, will occur on the borders of states, and these borders themselves will disappear in many places, prefiguring the day when they will cease to exist everywhere. (1905/2013, 219)

Calling this an 'era of mutual aid', Reclus normatively called for the dismantling of borders as part of a larger post-state and post-capitalism project. Effectively, the principal goal of this chapter closely tracks Reclus' vision of a world without borders as a means to overcome the inequalities and injustice within humankind. As Hyndman (2000) affirms, 'borders breed uneven geographies of power and status' (1). Pivoted on Bookchin's assertion that 'all our notions of dominating nature stem from the very real domination of human by human' (2001b, 13), a full realization of the dismantling of borders, of which constitute one of the most pernicious forms of violence against our sense of common humanity, would surely lead to a deeper appreciation of nature as our common home and one that must be nurtured.

Contrary to statist, neoliberal and sedentarist claims on the need for borders, human history is unarguably defined by a lack thereof. For example, Collyer (2014) notes that while the process of climate change has undoubtedly become a catalyst for environmental migration, it should also be recognized that 'for the entire history of human settlement, migration has been a form of adaptation to environmental stress' (117). Similarly, Boano et al. (2008) assert that

> of course forced displacement for environmental reasons is not a recent phenomenon. Scarcity of land resources and environmental degradation has led to waves of outmigration and/or conflict throughout history. Migration, and population movement in general, is part of human history and an important adaptive mechanism. (5)

And yet, migration and refugees have increasingly been treated as a problem, and largely statist, and sedentarist 'solutions' have served to reinforce 'cartographies of struggle' (Hyndman 2000, xvi) rather than provide harmony, dignity and freedom for the peoples of the world. Without respect for 'natural laws' – to quote Emma Goldman – whereby, for example, every human being has 'free access to the earth and full enjoyment of the necessities of life' (Goldman 1969, 62), 'man-made laws' have instead enabled the dominance of some over others, the destruction of the resources and splendour of nature, and the prevention – by means of barriers such as borders – of the natural process of adaptation.

Although a fundamentally anarchistic call for border transgressions, the growing body of post-colonial scholars has an interest and even responsibility in appreciating its value. This is because borders as an installment of colonialism and Western-styled asylum regimes as an installment of neocolonialism ought to be recognized as such. Effectively, 'the war of borders is a war waged by the West on a global scale to preserve its values' (Minh-ha 1991, 22), commingling with the history of capitalism which

> has been a series of attempts to solve the problem of worker mobility [. . .] since, if the system ever really came close to its own fantasy version of itself, in which workers were free to hire on and quit their work wherever and whenever they wanted, the entire system would collapse. (Graeber 2004, 61)

The process of globalization – for all of the violence and oppression it embodies through 'free' trade and imperialism – has also played a liberating role, undermining the relevance of borders, facilitating the widespread emergence of multicultural communities and enabling transnational networks of solidarity to form. These facts, although having largely served to the advantage of state and capital, can (and should) be used to deconstruct the authority and power of these profiteers for the benefit of all. Equally, they can be considered as signposts towards 'alter-globalization' (Pickerill and Chatterton 2006) and realizing Reclus' imaginary:

The spontaneous union across borders of men of good will remove all authority from certain falsely named 'laws' that were generalized from previous historical evolution and now that deserve to be relegated to the past as having had only relative truth. (1905/2013, 221)

As with Reclus and Kropotkin's notions of cooperation, freedom and harmony between our planet and its peoples, it is 'anarchist geographies' that inform us of the value and possibilities of a life without borders, understood as

> kaleidoscopic spatialities that allow for multiple, nonhierarchical, and protean connections between autonomous entities, wherein solidarities, bonds, and affinities are voluntarily assembled in opposition to and free from the presence of sovereign violence, predetermined norms, and assigned categories of belonging. (Springer 2012, 1607)

As political ecology is 'predicated on the assumption that any tug on the strands of the global web of human-environment linkages reverberates throughout the system as a whole' (Robbins 2012, 13), the case of environmental refugees exemplifies – as a reverberation – how the statist and capitalist global web and its linkages are effectively power-laden, antisocial and anti-environment. Thus, as confirmed by Simon Springer (2012), it is anarchist geographies that can provide an 'alternative geographical imagination' which can creatively reorganize and redefine the global web and its linkages for a better future for all inhabitants of planet Earth (Springer 2012). Part of this comprises imagining a world without borders between people and places since, as is accurately presented in the words of Bookchin (2011), 'frontiers have no place on the map of the planet, any more than they have a place on the landscape of the mind' (Bookchin 2001a, 25).

NOTES

1. For an outline of the emergence of the field of risk assessment and management, see Horlick-Jones (2013)

2. For a case study pertaining to the European Union–Turkey deal of 2016, see Parent (2016).

3. See, for example, Alarmphone (2015).

4. For Greece-Turkey border, see Parent (2016); for Bulgaria-Turkey border, see UN News Centre (2016),

5. Information based on multiple 'Governance Reports', available on Frontex' website, http://frontex.europa.eu/about-frontex/governance-documents/.

6. Ljubinka Brashnarska (senior external relations assistant, UNHCR Skopje), in discussion with the author, November 2016.

7. Data based on information available on 'UNHCR statistics', available on UNHCR website, http://popstats.unhcr.org/en/overview.

8. See, for example, Glantz (1987); Meadows et al. (1972).

9. For examples, see Lee (2001); Homer-Dixon (1991).

10. For an extensive analysis on the widespread use of blame avoidance in the Global North, see Hood (2010).

11. For a brief overview of the shift from dependency to neoliberalism as a development strategy, see Owusu (2003).

12. Examples of this include the British Raj (1858–1947), the Scramble for Africa (1881–1914) via the Berlin Conference of 1884–1885, French Indochina (1887–1954) and the territorial divisions in Asia Minor as a result of the Sykes-Picot Agreement (1915-17).

13. A good example of this is that of South America, where the development of borders was largely the result of nearly 100 years of nationalist independence conflicts across the continent, see Chapter 12 in Henderson (2013)

14. For a good overview of the relationship between Kropotkin and Reclus, see Ferretti (2011).

15. For a description of this, see 'Anarchism: Its philosophy and ideal' in *Kropotkin's Revolutionary Pamphlets*, ed. Roger N. Baldwin (New York: Dover Publications, Inc., 1970): 120–122.

REFERENCES

Akram, S. M. (2015). *Protecting Syrian Refugees: Laws, Policies, and Global Responsibility Sharing*. Prepared for Boston University International Human Rights Clinic. https://www.bu.edu/law/files/2015/08/syrianrefugees.pdf.

Alarmphone. (2015). *Moving On: One Year Alarmphone*. Berlin: Watch the Med.

Amnesty International. (2013). *Frontier Europe: Human Rights Abuses on Greece's Border with Turkey*. London: Amnesty International.

Baker, P., and McEnery, T. (2005). A corpus-based approach to discourses of refugees and asylum seekers in UN and newspaper texts. *Journal of Language and Politics* 4, no. 2: 197–226.

Barnett, M. (2014). Refugees and humanitarianism. In *The Oxford Handbook of Refugee and Forced Migration Studies*, edited by E. Fiddian-Qasmiyeh, G. Loescher, K. Long, and N. Sigona, 246. Oxford: Oxford University Press.

Beck, U. (1992). *Risk Society: Towards a New Modernity*. Thousand Oaks: SAGE Publications Ltd.

Black, R. (2001). *Environmental refugees: Myth or reality?* Working Paper No. 34, New Issues in Refugee Research, University of Sussex.

Black, R., Adger, W. N., Arnell, N. W., Dercon, S., Geddes, A., and Thomas, D. S. G. (2011). The effect of environmental change on human migration. *Global Environmental Change* 21, suppl. 1: s3–s11.

Boano, C., Zetter, R., and Morris, T. (2008). *Environmentally Displaced People: Understanding the Linkages Between Environmental Change, Livelihoods and Forced Migration*. Oxford: Refugee Studies Centre.

Bookchin, M. (2001a). Nationalism and the 'national question.' In *Free Cities: Communalism and the Left*, edited by E. Eiglad. London: Pluto Press.

———. (2001b). The ecological crisis and the need to remake society. In *Free Cities: Communalism and the Left*, edited by E. Eiglad. London: Pluto Press.

Bristow, S. D. (2007). *The political ecology of environmental displacement and the United Nations' response to the challenge of environmental refugees*. Doctoral Thesis, University of Wales. https://core.ac.uk/download/pdf/1919000.pdf.

Brown, E., and Cloke, J. (2004). Neoliberal reform, governance and corruption in the South: Assessing the international anti-corruption crusade. *Antipode* 36, no. 2.

Bryant, R. L. (1992). Political ecology: An emerging research agenda in third world studies. *Political Geography* 11, no. 1: 13.

Bylander, M., and Reid, G. (2017). Criminalizing irregular migrant labor: Thailand's crackdown in context. *Migration Policy Institute.* https://www.migrationpolicy.org/article/crimi nalizing-irregular-migrant-labor-thailands-crackdown-context.

Cardy, W. F. (1994). Environment and forced migration: A review. *Fourth International Research and Advisory Panel Conference.* Sommerville College, University of Oxford.

Carr, M. (2016). *Fortress Europe: Dispatches from a Gated Continent.* New York: The New Press.

Carrera, S. (2016). *Implementation of EU Readmission Agreements: Identity, Determination Dilemmas and the Blurring of Rights.* Brussels: Springer Open.

Castles, S. (2002). Environmental change and forced migration: Making sense of the debate. Working Paper, New Issues in Refugee Research, No. 70. Geneva: UNHCR.

Collyer, M. (2014). Geographies of forced migration. In *The Oxford Handbook of Refugee and Forced Migration Studies*, edited by E. Fiddian-Qasmiyeh, G. Loescher, K. Long, and N. Sigona, 112–126. Oxford: Oxford University Press.

Conisbee, M., and Simms, A. (2003). *Environmental Refugees: The Case for Recognition.* London: New Economics Foundation.

Davies, A., and Jacobsen, K. (2009). Profiling urban IDPs. *In Forced Migration Review FMR* 33: 13–15.

Davis, E. W. (2013). Beginning a sketch of accumulation by dispossession in contemporary Cambodia. *Cultural Anthropology*, 3. https://culanth.org/fieldsights/343-beginning-a-sk etch-of-accumulation-by-dispossession-in-contemporary-cambodia.

Dun, O., and Gemenne, F. (2008). Defining 'environmental refugee.' *Forced Migration Review FMR* 31: 10.

Ferretti, F. (2011). The correspondence between Élisée Reclus and Pëtr Kropotkin as a source for the history of geography. *Journal of Historical Geography* 37, no. 2: 216–222.

Gabrielli, L. (2014). Securitization of migration and human rights: Frictions at the southern EU borders and beyond. *Urban People Lide Mesta* 16, no. 2: 311–322.

Gadgil, M., and Guha, R. (1995). *Ecology and Equity: The Use and Abuse of Nature in Contemporary India.* Abingdon: Routledge.

Glantz, M. H. (Ed.). (1987). *Drought and Hunger in Africa.* London: Cambridge University Press.

Goldman, E. (1969). Anarchism: What it really stands for. In *Anarchism and Other Essays*, edited by R. Drinnon, 62. New York: Dover Press.

Goodwin-Gill, G. S. (2001). After the Cold War: Asylum and the refugee concept move on. *Forced Migration Review FMR* 10: 14.

Graeber, D. (2004). *Fragments of an Anarchist Anthropology.* Chicago: Prickly Paradigm Press.

Greenberg, J. B., and Park, T. K. (1997). Political ecology. *Journal of Political Ecology* 1, no. 1: 1–12.

Hall, M. (2011). Beyond the human: Extending ecological anarchism. *Environmental Politics* 20, no. 3: 374–390.

Hammer, T. (2004). Desertification and migration: A political ecology of environmental migration in West Africa. In *Environmental Change and Its Implications for Population Migration*, edited by J. D. Unruh, Maartens, S. Krol, and Nurit Kliot, 231–245. Dordrecht: Kluwer Academic Publishers.

Hansen, R. (2014). State controls: Borders, refugees and citizenship. In *The Oxford Handbook of Refugee and Forced Migration Studies*, edited by E. Fiddian-Qasmiyeh, G. Loescher, K. Long, and N. Sigona, 256. Oxford: Oxford University Press.

Haselberger, B. (2014). Decoding borders. Appreciating border impacts on space and people. *Planning Theory* and *Practice* 15, no. 4: 505–526.

Henderson, P. V. N. (2013). *The Course of Andean History*. Albuquerque: University of New Mexico Press.

Homer-Dixon, T. (1991). On the threshold: Environmental changes as cause of acute conflict. *International Security* 16, no. 2: 76–116.

Hood, C. (2010). *The Blame Game: Spin, Bureaucracy and Self-Preservation in Government*. Princeton, NJ: Princeton University Press.

Horlick-Jones, T. (2013). Risk and time: From existential anxiety to post-enlightenment fantasy. *Health, Risk & Society* 15, no. 6–7: 489–493.

Human Rights Watch. (2017, March 15). Greece: A year of suffering for asylum seekers. *HRW News Release*. https://www.hrw.org/news/2017/03/15/greece-year-suffering-asylum-seekers.

Humphrey, M. (2014). Securitization of migration: An Australian case study of global trends. *Revista Latinoamericana de Estudios sobre Cuerpos, Emociones y Sociedad* 6, no. 15: 83–98.

Hyndman, J. (2000). *Managing Displacement: Refugees and the Politics of Humanitarianism*. Minneapolis: University of Minnesota Press.

International Federation of Red Cross and Red Crescent Societies. (2009). *Climate Change and Human Mobility: A Humanitarian Point of View*. Geneva: International Federation of Red Cross and Red Crescent Societies.

Jacobson, J. (1998). *Environmental Refugees: A Yardstick of Habitability*. Washington: World Watch Institute.

Kainz, L. (2016). People can't flood, flow or stream: Diverting dominant media discourses on migration. *Border Criminologies*. https://www.law.ox.ac.uk/research-subject-groups/centre-criminology/centreborder-criminologies/blog/2016/02/people-can't.

Kälin, W. (2014). Internal displacement. In *The Oxford Handbook of Refugee and Forced Migration Studies*, edited by E. Fiddian-Qasmiyeh, G. Loescher, K. Long, and N. Sigona, 163–175. Oxford: Oxford University Press.

Keane, D. (2004). The environmental causes and consequences of migration: A search for the meaning of environmental refugees. *Georgetown International Environmental Law Review* 16, no. 1: 209–223.

Kibreab, G. (1997). Environmental causes and impact of refugee movements: A critique of the current debate. *Disasters* 21, no. 1: 20–38.

Kinna, R. (1995). Kropotkin's theory of mutual aid in historical context. *International Review of Social History* 40, no. 2: 259–283.

Kropotkin, P. (1902/1955). *Mutual Aid: A Factor of Evolution*. Boston: Extending Horizons.
———. (1970). *Kropotkin's Revolutionary Pamphlets*, edited by R. N. Baldwin. New York: Dover Publications, Inc.

Lazaridis, G., and Wadia, K. (Eds.). (2015). *The Securitization of Migration in the EU: Debates Since 9/11*. London: Palgrave Macmillan UK.

Lebaron, G., and Ayers, A. J. (2013). The rise of a 'new slavery'? Understanding African unfree labour through neoliberalism. *Third World Quarterly* 34, no. 5: 873–892.

Lee, S. (2001). Emerging threats to international security: Environment, refugees, and conflict. *Journal of International and Area Studies* 8, no. 1: 73–90.

Leghtas, I. (2017). *"Like a Prison": Asylum-Seekers Confined to the Greek Islands*. Washington: Refugee International.

Loescher, G. (2014). UNHCR and forced migration. In *The Oxford Handbook of Refugee and Forced Migration Studies*, edited by E. Fiddian-Qasmiyeh, G. Loescher, K. Long, and N. Sigona, 219. Oxford: Oxford University Press.

Long, K. (2013). When refugees stopped being migrants: Movement, labour and humanitarian protection. *Migration Studies* 1, no. 1: 4–26.

Martin, L. (2011). Is Europe turning into a 'technological fortress'? Innovation and technology for the management of EU's external borders: Reflections on FRONTEX. In *Regulating Technological Innovation: A Multidisciplinary Approach*, edited by M. A. Heldeweg and E. Kica, 131–151. London: Palgrave Macmillan UK.

McKay, F. (2013). A return to the 'Pacific Solution. *Forced Migration Review* 44: 24–26.

McNamara, K. E. (2008). Pragmatic discourses and alternative resistance: Responses to climate change in the Pacific. *Graduate Journal of Asia-Pacific Studies* 6, no. 2: 33–54.

Meadows, D. H., Meadows, D. L., Randers, J., and Behrens, W. W., III. (1972). *The Limits to Growth: A Report for the Club of Rome's Project on the Predicament of Mankind*. London: Earth Island.

Médecins Sans Frontières (2014). *Invisible Suffering: Prolonged and Systematic Detention of Migrants and Asylum Seekers in Substandard Conditions in Greece*. Geneva: Médecins Sans Frontières.

Migreurop. (2012). Deportation from European territory 'At Any Cost'. *News Release No. 1*. http://www.migreurop.org/IMG/pdf/note_de_migreurop_062012_accords_de_r_r_ad mission_version_anglaise_web_version.pdf.

Minh-ha, T. T. (1991). *When the Moon Waxes Red: Representation, Gender and Cultural Politics*. Abingdon: Routledge.

Morrissey, J. (2012). Rethinking the 'debate on environmental refugees': From 'maximalists and minimalists' to 'proponents and critics.' *Journal of Political Ecology* 19, no. 1: 36–49.

Muedin, A. (2010). Readmission as a mechanism of general return management process. Presentation International Organization for Migration. www.unitar.org/ny/sites/unitar.org.ny/files/4-_Readmission_Agreements_.ppt.

Myers, N. (2002). Environmental refugees: A growing phenomenon of the 21st century. *Philosophical Transactions: Biological Sciences* 357, no. 1420: 609–613.

Myers, N., and Kent, J. (1995). *Environmental Exodus: An Emergent Crisis in the Global Arena*. Washington: Climate Institute.

Neumann, R. P. (2009). Political ecology: Theorizing scale. *Progress in Human Geography* 33, no. 3: 398–406.

Otunnu, O. (1992). Environmental refugees in sub-Saharan Africa: Causes and effects. *Refuge* 12, no. 1: 13.

Owusu, F. (2003). Pragmatism and the gradual shift from dependency to neoliberalism: The World Bank, African leaders and development policy in Africa. *World Development* 31, no. 10: 1655–1672.

Padoan, L. (1995, November 30). COP 21: Climate change, refugees and couture. *UNHCR Press Release*. http://www.unhcr.org/565c13c26.html.

Parent, N. (2016, September 27). Legitimization of uncertainty: The shaky fate of Syrian migrants in Turkey and Europe. Lecture, *Centre for Refugee Studies Seminar Series*, York University, Toronto, Canada. https://nikparent.files.wordpress.com/2016/10/seminar_l egitmization-of-uncertainty.pdf.

Parker, S. (2015). 'Unwanted invaders': The representation of refugees and asylum seekers in the UK and Australian print media. *eSharp*, 23: 1–21.

Pickerill, J., and Chatterton, P. (2006). Notes towards autonomous geographies: Creation, resistance and self-management as survival tactics. *Progress in Human Geography* 30, no. 6: 730–746.

Pickering, S. (2001). Common sense and original deviancy: News discourses and asylum seekers in Australia. *Journal of Refugee Studies* 14, no. 2: 169–186.

Pro Asyl. (2013). *Pushed Back: Systematic Human Rights Violations Against Refugees in the Aegean Sea and at the Greek-Turkish Land Border*. Frankfurt: Pro Asyl.

Reclus, E. (1894). L'Anarchie. Presented at *Les Amis Philanthropes'*. Brussels, Belgium.

———. (1905). *L'Homme et la Terre* (Vol. 1). (trans) E. Reclus. Paris: Librairie Universelle.

———. (1905/2004). Progress. In *Anarchy, Geography, Modernity: The Radical Social Thought of Elisée Reclus*, edited by J. P. Clark and C. Martin, 234. Lanham: Lexington Books.

———. (1905/2013). Progress. In *Anarchy, Geography and Modernity*, edited by J. P. Clark and C. Martin, 219. Oakland: PM Press.

———. (1995). *Man and Nature*. Petersham: Jura Media.

Renfrew, D. (2011). The curse of wealth: Political ecologies of Latin American neoliberalism. *Geography Compass* 5, no. 8: 581–594.

Rimmer, C. (2017). *EU Migration Policy and Returns: Case Study on Afghanistan*. Brussels: European Council on Refugees and Exiles.

Robbins, P. (2012). *Political Ecology*. New York: John Wiley and Sons Ltd.

Rodier, C. (2013). The externalisation of migration controls. In *Shifting Borders: Externalising Migrants Vulnerabilities and Rights?* Brussels: International Federation of Red Cross and Red Crescent Societies.

Scurrah, N., and Hirsch, P. (2015). *The Political Economy of Land Governance*. Vientiane: Mekong Region Land Governance.

Sims, K. (2015). The Asian Development Bank and the production of poverty: Neoliberalism, technocratic modernization and land dispossession in the Greater Mekong Subregion. *Singapore Journal of Tropical Geography* 36, no. 1: 112–126.

Springer, S. (2012). Anarchism! What geography still ought to be. *Antipode* 44, no. 5: 1605–1624.

Springer, S., and Inca, A. (2012). Reanimating anarchist geographies: A new burst of colour. *Antipode* 44, no. 5: 1591–1604.

Suhrke, A. (1994). Environmental degradation and population flows. *Journal of International Affairs* 47, no. 2: 47–48.

Tobias, S. (2012). Neoliberal globalization and the politics of migration in sub-Saharan Africa. *Journal of International and Global Studies* 4, no. 1: 1–16.

Trombetta, M. J. (2014). Linking climate-induced migration and security within the EU: insights from the securitization debate. *Critical Studies on Security* 2, no. 2: 131–147.

UNHCR Regional Representation in Canberra. (2017, August 8). UNHCR warns of escalating crisis on Manus Island. *UNHCR Press Release*. http://www.unhcr.org/news/press/2017/8/598909c17/unhcr-warns-of-escalating-crisis-on-manus-island.html.

United Nations. (2016, August 11). UN rights chief sees 'worrying signs' in Bulgaria's detention regime for migrants. *UN News Centre*. http://www.un.org/apps/news/story.asp?NewsID=54664#.WaWqzD6GOM9.

Walker, P. A. (2006). Political ecology: Where is the policy? *Progress in Human Geography* 30, no. 3: 382–395.

Walsh, C. (2014). Bordered spaces: Nation-States and private property. *Public Political Ecology Lab*. http://ppel.arizona.edu/?p=684.

Yonetani, M. (2015). *Global Estimates 2015: People Displaced by Disasters*. Geneva: Internal Displacement Monitoring Centre.

Young, J. G. (2017). Making America 1920 again? Nativism and US immigration, past and present. *Journal on Migration and Human Security* 5, no. 1: 217–235.

Zetter, R., and Morrissey, J. (2014). The environment-mobility nexus: Reconceptualizing the links between environmental stress, (im)mobility, and power. In *The Oxford Handbook of Refugee and Forced Migration Studies*, edited by E. Fiddian-Qasmiyeh, G. Loescher, K. Long, and N. Sigona, 343. Oxford: Oxford University Press.

4

Questioning Capitalistic Power Structures

A Way to Reconnect People with Lands?

Simon Maraud and Etienne Delay

The idea of writing a chapter on the relationship between stakeholders and their land, and the obstacles put in place in order to limit this relationship, emerged during various discussions on our different works (Delay and Linton 2018; Maraud and Desbiens 2017; Linton and Delay 2018). We became very interested when we realized that similar issues appeared in two very different contexts, our two PhD research fields. On seeing these connections, we felt it quite important to understand how comparable mechanisms could take place in the South of France with farmers and in northern Sweden with the Sami people. Anarchism came into the project later on.

In this book, which focuses on the role of the state in the management of energy and nature, we have chosen to emphasize the empiric knowledge about the state, rather than the theoretical. The theory on the state is very important but the empiric method makes it more material (Bourdieu 1990 in 2012). It does not mean that theory on the state is useless, quite the contrary, but that it is not our purpose here. The aim of this chapter is to analyse how – in our two cases – state structures are creating a gap between stakeholders and the land, but not to construct a universal rule about this gap. We thus follow a very empirical method that is based on our field research, making our results highly contextualized. However, this chapter also discusses general issues with the creation of a framework to grasp the context of domination regarding the societal organization of groups in the context of management of natural resources. This framework would thus be applicable in other contexts. After presenting our two case studies, we shall discuss what we call the TerritOry-Resou rces-Societal-Organization (TORSO) framework. Following this, we shall discuss the results arising from our examples and from the framework itself.

Irrigation in Southern France

The Roussillon plain, in the Pyrénées-Orientales Department of France, is an ideal area in which to study the collective management of water resources. The region has a long history of community management of irrigation and some momentous recent changes in the physical availability of water have had an impact on management practices. Located in the Mediterranean region, along the border with Spain, the hydrological and climatic conditions have played an important role in the emergence of historical irrigation networks (Jaubert de Passa 1821: 261), which remain partially in place today, along with the local economy that is so closely tied to them. Today, as in the past in this region, water remains inseparable from agriculture and agriculture remains inseparable from water (Ruf 2001).

We are looking specifically at the Têt bassin, within the Roussillon plain, in which there are two main dams: the Bouillouse dam, built in the first decade of the twentieth century, and the Vinça dam, constructed in the 1970s. This study takes place in the area between the Vinça dam to the east, the Corbières mountain chain to the north, the Mediterranean Sea to the west and the Albert mountains to the south, focusing on the basin of the Têt River.

The Mediterranean climatic conditions could be said to determine the necessity of irrigation.[1] Here, producers have long adopted local, decentralized structures to manage irrigation. Jaubert de Passa (1821) identifies these decentralized hydrological management structures as typical of border regions that have been affected by historical invasions and political uncertainty, in the sense that the influence of the state has been historically weak in this region.

The World Heritage Site of Laponia

Laponia is a UNESCO World Heritage Site in Sweden, chosen for its cultural and natural values. It comprises four national parks and three nature reserves. Laponia is thus quite a symbolic area for various reasons. First of all, it is in Sápmi, the land of the Sami people, and this has been recognized as part of the heritage to be emphasized by UNESCO. The Sápmi is located in the northern parts of Norway, Sweden and Finland and in the Kola Peninsula in Russia. The Sami are the indigenous people of Northern Europe; they number approximately 80,000 in total, 20,000 of which are in Sweden.[2] Another reason is that two of the Laponia national parks were the first in Sweden, and indeed in Europe. *Sarek* and *Stora Sjöfallet* were created in 1909, followed by *Muddus* in 1942 and *Padjelanta* in 1962. This is therefore a place with high conservationist values. Finally, Laponia is a symbolic space because it represents one of the only large-scale victories for the Sami in Sweden. The *Laponiatjuottjudus* – or Laponia Process – started in 1996, when UNESCO recognized the cultural and natural values of the site. The site was already protected by the state, but UNESCO encouraged the government of Sweden to involve the Sami people in the board of management. Since the Sami – or more precisely the Sami reindeer herders[3] – used the place with their herds, UNESCO considered it appropriate to make them part of Laponia. The struggle to include the Sami in management decisions lasted fifteen years before a real change happened, and the Sami now see this as a form of decolonization. This site therefore presents some very particular challenges and trajectories in terms of emancipation of the indigenous Arctic Indigenous population. The area contains and represents a number of divergent stakeholders – and thus interests – in the same territory. However, Sweden has imposed its own state territoriality, so the area is not perceived as being decolonized, and as such it is appropriate to investigate this new structure and to identify the stakeholders who are making decisions about the land.

METHODS – TORSO

As described in the previous section, our two case studies are quite distinct, as well as being geographically distant from each other. In both cases, however, the current conflicts between the states and the local stakeholders highlight the local socio-spatial history. In order to make a comparison between water management practices in southern France and the Laponia board of management in northern Sweden, we have developed a framework that relates to the ways in which people interact collectively with natural resources.

A Framework for Domination Processes

In geography, analysing the dynamic relationship between the human and non-human (in this case resources) and among humans themselves is a burning issue. As geographers, we have worked in different contexts, from the Arctic to the Sahel.[4] In all of the situations we studied, we focus on how humans are using local resources, and how exogenous stakeholders – especially the state – interfere in this process in order to sustain power. Identifying what seemed to be a similar mechanism, we asked ourselves if it might be possible to create a framework to represent the social dynamics of human groups regarding natural resources management and its spatial and temporal evolution.

Political ecology studies 'ecological distribution' and the use of power in order to gain access to natural resources (Martinez Alier 2002). This 'ecological distribution' creates conflicts, which are studied in political ecology in order to establish the domination processes involved in access to, control of and dependence on natural resources (Martinez Alier 2002). Political ecology allows us to work on the relationships between societies and nature from a power perspective (Escobar 1996). Arturo Escobar (1996) explains that such an angle implies work in other domains as political economy and its discursive and practical productions. An anarchist political ecology would propose new narratives on the politics of resources management. It begins with how to work on an issue. Agnew (1994) demonstrates that the national scale is not the only relevant one when working on the state. Indeed, the scale we choose is not apolitical; it has been socially constructed and refers to an established order (Leitner and Miller 2007). To refuse a scale-based approach to a phenomenon at a national level would make no sense. However, it is essential to be conscious of this bias in the study. Only then can spaces be reshaped and made dynamic to create a new scalar configuration (Green 2016). To work on an alternative system implies rethinking the scales we use as researchers (Neumann 2009).

We called this framework TORSO, which means TerritOry-Resources-Socie tal-Organization. It seemed to us that these four notions were the keywords for our project. We consider them to be completely interrelated. The overall idea of this framework is to demonstrate that the relationship between societies and the environment depends on the relationships among humans (Bookchin 1971). In other words, it is about highlighting the domination processes that take place in the development of management systems. By understanding the hierarchy involved, along with the disparity in the importance of various stakeholders' values in resources management, we think that TORSO could be used as a step to repair the damage caused by a capitalistic neoliberal society. If the state, or another exogenous stakeholder, creates a gap between local people and their land, the land then becomes a tool to make people dependent on the authority, which tends to prevent them from questioning the power structure/their power.

The TORSO framework studies the transitional aspect of different management attitudes, in particular the evolution of the stakeholders' position and the position of the structure itself. Without this dynamism, we would be likely to see an institutionalization of the structure and then a return to hierarchies among stakeholders. Of course, hierarchies are acceptable as a temporary situation during transition (Kropotkin 1892). If the buzz of excitement about formalizing such a new management structure fades away, the organizers within the hierarchy may be tempted to try to maintain the status quo. In such cases, the necessary changes must be made; hierarchies should not be allowed to continue. In a way, TORSO is quite close to the *social transformism* of De Greef (1895). In his book on progress and regress, De Greef studied the transitional forms of universal structures, laws and rules. Knowledge regarding these aspects, of what fundamentally makes a society what it is, demonstrates that this can result in progress or regress. Progress is seen as an evolution, which increases and lasts (De Greef 1895). TORSO is also about just one field of *social transformism*: natural resources management. However, it can be used in a similar manner to analyse the evolution of common values and rules about a collective way of living. Taking into account this evolving dynamic implies the study of a human group over a period of time. In this respect, TORSO is a framework that can be applicable at various spatial and temporal scales. For this reason, it is applicable to our two case studies of water management in the South of France and the evolution of its management from the 1200s to today at a regional scale, as well as the Laponia resources from 1996 to today at the scale of a protected area. In considering the various scales of analysis, a TORSO framework gives a holistic view of the processes of domination and power that relate to situations in specific territories but which have a basis in common, such as the confrontation between a 'social form' and an 'anarchic-gregarious' way of collective living, both of which will be discussed in the following section (MacDonald 2016).

Organized Society and Anarchic-Gregarious Living

Analysis of the role of the state, or other external stakeholders (international companies, for example), is one of the bases of the anarchist approach, and to understand the spatial impact of it is a method used in geography. This dual process of anarchism and geography is also reproduced within the TORSO framework.

The anthropologist Charles MacDonald (2016, 2014, 2013, 2011) makes a comparison between two types of collective living for human beings. For him, there is, on the one side, the 'social form' of collective living, and on the other side, the 'anarchic-gregarious' type. Other authors would use the opposition between 'authority' and 'liberty', which is a very similar approach, but we find that of MacDonald to be more holistic.

According to MacDonald, the social form – or 'social-hierarchic' or 'organized society' (MacDonald 2016) – is the most common way of living collectively, especially as a result of colonialism. This characterizes the social organization of a society, with classes, hierarchies and thus inequalities. The social form of collective living always implies power (MacDonald 2014).

The anarchic-gregarious form of collective living – or 'open aggregated' (Gibson and Sillander 2011) – is less common today, and is never presented as a possible type of organization by people in power, even if *anarchs*[5] still exist in certain marginalized places (MacDonald 2011). Even without this formal acknowledgement, we do think that an anarchist attitude must be taken into account during the creation of a territory, since it supposes an organizational form where there is no hierarchy, no chief and where all members of the group are equal.

Kropotkin (1921) made a similar comparison between two tendencies in humanity, the first regarding the seizure of power in order to dominate, and the second to resist and to maintain equality (Pelletier 2013). According to Kropotkin (1901), these are the statists and the anarchists.

With MacDonald's addition to our framework, the idea is to study the possibilities of cooperation and solidarity between the stakeholders in the management of land. As such, the TORSO framework showcases the importance of partnership, which is a fundamental point for the autonomy of people (MacDonald 2014). Making people less autonomous is possible (MacDonald 2016), and this is actually a useful tool for depoliticizing a stakeholder with land. By creating a gap between these two entities, it is possible to implement a position dependent on the established power (Bookchin 1990; 1971). Autonomy is thus essential to remove competing rules and to create a common goal (Pelletier 2013; Kropotkin 1892; Metchnikoff 1886).

However, this common goal must not be transcendent – with religion, state, Nation and so on – and the objective has to be a plural society without domination

(MacDonald 2016). Only then does it become progress. The framework is organized with:

- The abscissa axis based on the degree of intensity of attitudes to tend to an anarchic-gregarious way of collective living.
- The ordinate axis based on the degree of intensity of attitudes to tend to a social form of collective living.

With this conceptual diagram, it is possible to position a specific social group depending on its way of managing a natural resource. The evolution of the management structure must be studied over several time periods. By comparing the various degrees of collective attitudes, it is possible to position the group between 'Social' and 'Anarchic-gregarious' patterns. To implement an egalitarian society is a very long process that includes the formulation of complex rules (MacDonald 2014). It takes into account twelve criteria in different territories in order to understand domination mechanisms, attempts at emancipation, failures and revolutions.

TORSO is not a statistic approach, but a qualitative one; it is not a general theory but a framework, based on anarchist theories, to grasp different territories with twelve criteria to understand domination mechanisms, emancipatory attempts, failures and revolutions.

Criteria for Collective Attitudes

In TORSO, the position of a society depends on twelve different criteria. Each criterion allows us to qualify the attitude of a group on a scale from the social form to the anarchic-gregarious form of collective living. Depending on the degree of intensity of the variable, we get a -1, -0.5, 0, $+0.5$ or $+1$. The accumulation of points at the end of the table gives the position of the society on the diagram for a specific time period. The distribution of the criteria defines the strategic questions regarding each trajectory (figure 4.1).

We decided to not make a hierarchy of these criteria because we think that each one of them makes the society closer to a type of collective living, from the closest trajectory to the social form to that closest to the anarchic-gregarious form of collective living, and that each one of them is interrelated.

Management Practices

The management practices covers situations from the *associative profile* ($+1$) to the *representative profile* (-1). The anarchic-gregarious profile directly tends to a very robust associative life. By 'associative life' we are referring to Kropotkin's (1892: 117–119) idea, which is a societal organization form divided into common and individual tasks. A set of associations offers diverse services to the group in order to satisfy its needs by giving capable people a chance to participate. If someone is

Society / Time	Anarchic-gregarious form of collective living +1	+0.5	0	-0.5	Social form of collective living -1
Management practices	Associative profile				Representative profile
Spatial scale	Ultra-local				Supra-national
Demographic scale	Low amount of inhabitants				High amount of inhabitants
Bureaucratic significance	Low				High
Accessibility	Including				Excluding
Dependence on market	Low				High
Property form	Low				High
Authority	Nonexistent				High
Consensus significance	High				Low
Dependence on resources	High				Low
Impact ratio on resources	High				Low
Structure morphology	Multipolar				Centralized
Total	12	6	0	-6	-12

Figure 4.1   Table of the TORSO Criteria.

concerned about a certain number of issues, he or she will be able to participate in different associations to deal with them. The rest of the time, the individual will have time to commit to recreational activities (Kropotkin 1892). This system, as a combination of multiple associations, enables people to avoid a situation where there is a monopoly of competences within only one institution. The representative profile concerns societies that delegate managing powers to other individuals presented as having the skills to deal with distinct topics on behalf of everyone. So, if a person is concerned by a certain amount of issues, she or he will be able to elect – or not – a representative to deal with them along with other elected representatives. In other words, there is a monopoly of competences, which can be targeted towards social needs, race and gender for instance.

Spatial Scale

The spatial scale of TORSO embraces situations from *ultra-local* (+1) to *supra-national* (–1) profiles. This scalar diversity is paramount for the effective operation of the previous criterion. Indeed, the smaller the scale, the greater the gap will be between the representatives and the represented, and vice versa. So, the ultra-local scale easily allows the implementation of an associative profile due to the proximity between the individuals of the group and with the land in question. On the contrary, we think that the greater the territory, the less likely it is that the decision-makers in the management structure will be close to it. For example, if the management scale is supra-national, it becomes quite difficult to avoid a social form of collective living. Indeed, when individuals make decisions about territories in which they don't practice, they are not affected by the impact of their decisions.

Demographic Scale

The demographic scale incorporates situations from a *low number of inhabitants* (about a hundred) (+1) to a *high amount of inhabitants* (above a million) (–1). Similar to the spatial scale, the demographic scale is an essential criterion for establishing a collective living system. This is because the larger the group, the more distance there will be between individuals. Alternatively, the smaller the group is, the higher the chance of a successful implementation of a local network of associations. Murray Bookchin (1971) demonstrates that with 'libertarian municipalism', this type of group can network with other groups. It means that a limited group does not mean living in autarchy but being able to form local relationships with others nearby. However, if the group comprises a million people, they will not be able to lead a collective life without hierarchic forms. Thus there will be more opportunities to establish a representative system to take decisions, and some individuals will be able to be decision-makers in a normative environment, while others will not.

Bureaucratic Significance

Bureaucratic significance comprises situations from a *low bureaucratic significance* (+1) to a *high bureaucratic significance* (–1). Bureaucracy is one of the most important tools of the social form of collective living. It is closely related to the precedence criteria. It embodies the idea of delegating decision-making power and skills relating to various problems to so-called specialized institutions.[6] The more powerful the bureaucracy, the greater is the gap between the individuals in a group and their territory. Since the skills of the bureaucracy are specific to each body of the structure and professionalized, any stakeholder who is not familiar with the mechanism and the modalities of the institutions is not able to act in the decision-making process. Alternatively, the weaker the bureaucracy, the more likely it is that individuals will have the chance to use their empirical knowledge and practice, their vision and their needs in making decisions regarding their land.

Accessibility

The accessibility criterion covers situations from *including profile* (+1) to *excluding profile* (–1). It allows us to understand to what extent a management structure is closed or open to different stakeholders of the territory. We consider an including profile as an association within which anyone interested in participating in the collectivity is legitimately able to do so. An exclusive structure means that stakeholders with a range of skills, ideologies, diplomas, social classes, races, genders and ages are involved in the management system. Thus, in an excluding structure, a hierarchy of legitimacy is possible.

Dependence on Market

The dependence on market encompasses situations from a *low dependence* (+1) to a *high dependence* (–1) on market. A score of high dependence on market implies decisions towards this dependency, which is a reflection on the strength of the neoliberal capitalism. Neoliberal governance creates dependency forms on the market and the competition, which leads to a system that commodifies a set of goods that encompass what the land and the society are. This gradual reliance distorts the decision-making process regarding management. A society that depends highly upon the market will have a social form of collective living because there will be competition between various capitals. On the other hand, if a group of individuals has a little direct dependence on the market, we think that its freedom of decision-making processes will be greater in terms of management. It will thus be possible to take decisions rationally, without the pressure of economic growth.

Property Form

This criterion comprises situations from a *low form* (+1) to a *high form* (–1) *of ownership*. The idea is not to deny all types of property ownership. Proudhon (1840) says that ownership has been hijacked from its original meaning which is liberty, and not abuse. It is thus the abuse of property that we are evoking here. If, among different stakeholders all working in the territory, owning property makes one person more legitimate than another because he or she has more land, then a hierarchy exists. On the contrary, if the group of individuals working on the land can decide jointly on a common management, then the hierarchy tends to decrease.

Authority

The authority criterion incorporates situations from *nonexistent* (+1) to *high form* (–1) *of authority*. Authority concerns the imposition of governance. If it is high, there is a hierarchy in the society because one or some individuals have the power to establish their model of management. They have various tools available to maintain order around their governance, and they can use force if they consider it necessary, for example, via the police, the justice system or the administration. On the other hand, a model without authority means that the skills used are from the whole group of individuals without any imposition of order.

Consensus Significance

The consensus significance criterion covers situations from a *high significance* (+1) to a *low significance* (–1) *of consensus.* It is the logical follow-up to the authority criterion. During the decision-making step, it is important for the anarchic-gregarious system to be implemented that every member of the group has an equal say on diverse issues within the arena of discussion.

Every decision made should be the result of a consensus among all the stakeholders concerned by the decision itself. However, if the level of consensus is weak, or does not exist, there is no anarchic ambition but a social form of interaction since some stakeholders have a bigger say than others in the territory.

Dependence on Resources

The dependence on resources criterion embraces situations from *a high dependence* (+1) to a *low dependence* (–1) *on resources.* We consider that if a group has a high dependence on resources and on its habitat, it will be conscious of the importance of them. Capitalism is a doctrine implementing a gap between the humans and their territories, with control of resources, overconsumption and a belief in economic growth without limit (Bookchin 1990; Rist 2007). The importance of dependency on resources has a high impact on the management policies put in place. We think that if a group of individuals depends greatly on certain resources, it will try to rationalize its consumption of them in order to maintain them, since the community future is related to it (Springer 2016; Ingold 2009). However, if a group does not depend on the resources around it, or if it thinks the resources as unlimited, it will not be able to consider rational management. Instead, it might be tempted to enter into a market logic regarding them.

Impact Ratio on Resources

The impact on resources incorporates situations from *a high impact ratio* (+1) to a *low impact ratio* (–1) *on resources.* As regards dependency on resources, we think that if the way of life of a group, or that of another group, is highly impacted, it will try to minimize this impact in order to be affected as little as possible. So, a group that is highly impacted by human activities might try to introduce more rational practices on its land, since it can anticipate the level of repercussions that there will be on its future. On the other hand, a group that does not suffer any direct impact will have more difficulty in admitting its responsibility for these fuzzy impacts and in questioning its way of life.

Also, we think that a group that is having a high impact on its resources, but which is also suffering from this impact, will be better able to consider more appropriate management of the land, and vice versa. So, this criterion is a ratio between the impact on the resource itself and the effect of the resource having suffered such an impact has on a group's way of life.

Structure Morphology

Finally, structure morphology covers situations from a *multipolar profile* (+1) to a *centralized profile* (–1). The morphology of a management model is a very important analysis of the modalities regarding the societal system. A multipolar profile allows the importance of the bureaucracy to be reduced; the great institutions of governance give way to local decision-making arenas instead. Thus a decentralized form might facilitate a network between localities, and this without impacting the governance of any of them. However, a centralized model creates hierarchies not only between individuals but also between spaces. In other words, there will be powerful places – with headquarters, events and infrastructures that will legitimatize the centralization – and weak places, which become the subjects.

All these criteria give a 'range of attitudes' (Lacroix 1981), which is essential to define the position of a group at a particular time. Since the level of intensity of each variable fluctuates over time, the diagram shows a dynamic evolution of the global attitude of a group as regards its societal organization for the management of the resources of the territory in which it lives. Also, we voluntarily chose not to use a hierarchy of importance between the criteria; they all encourage changes of direction in the collective behaviour within the structure (Lacroix 1981).

RESULTS

We applied TORSO framework to our two examples in order to analyse the organization of a society as regards the land. Using a common base of criteria, the idea of TORSO is to use comparison in order to provide mutual inspiration for societies considering several, quite diverse, possible trajectories.

TORSO for the Eastern Pyrenees

Political power in this region has been decentralized since the invasion of the Moors in the Middle Ages. They brought with them techniques for water management and canal construction. Initially, local political and economic power was achieved by using the canals to provide water power for industrial activities (milling, textile production). The infrastructure was privately owned (seigneurs, abbeys, leaseholders) and the canals were managed and maintained by groups of property holders organized as associations, or syndicates (Ladki et al. 2012). Irrigation, although present in the first millennium, began growing in importance. Initially this growth in importance was due to water in the canals being available for a secondary use. Having begun in the twelfth century, it became the dominant use in the eighteenth century, as industry declined. By the nineteenth century, viticulture became predominant, and industrial use of water disappeared completely (Ladki et al. 2012).

There are five steps on the TORSO diagram (figure 4.2). The first one is for the early part of this period; the region (Catalonia) was attached to the Kingdom of Aragon and then became part of the Kingdom of Majorca from 1262 to 1344 before falling back into the hands of the Kingdom of Aragon. French Catalonia returned to the sphere of influence of the King of France in the seventeenth century. Throughout all these constitutional arrangements, the customs, laws and rights regarding the collective management of canals and irrigation remained fundamentally unchanged. They were

> confirmed by official declarations [. . .]. The drafting of the new Civil Code at the time of the French Revolution included in article 645 the introduction of respect . . . [and] maintenance of particular and local regulations on watercourses and the use of water. (Jaubert de Passa 1821: 261–262)

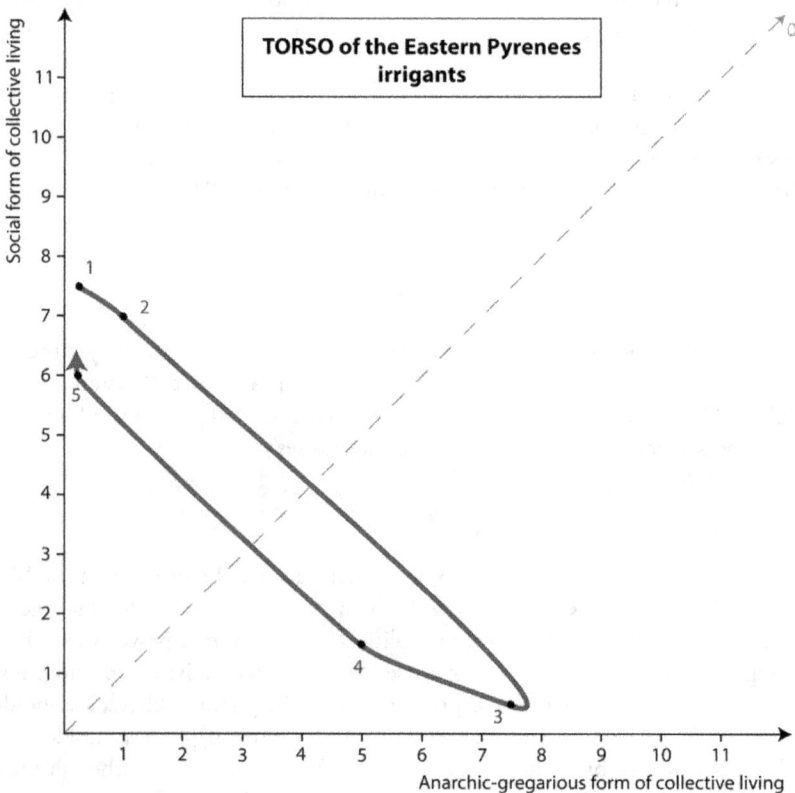

Realisation: S Maraud, E Delay, 2017

Figure 4.2   TORSO of Irrigation in the Eastern Pyrenees.

Thus, the pattern of collective ownership and management of water in the region withstood numerous changes of constitutional form. This basic rule of 'self-managed' and 'community' irrigation systems continued throughout the nineteenth century, through the various constitutional adventures and changes in the French state (Ruf 2009). The second point (figure 4.2) describes the law of 21 June 1865, when the state codified these arrangements, creating a legal entity of the Authorised Syndicated Association (ASA), conferring on them the status of public agency or institution (*établissement public*). This was followed by considerable expansion of gravity-based irrigation networks throughout the region (Ladki et al. 2012).

The next major expansion of irrigated agriculture in the region occurred immediately after the Second World War, as France was working full time to rebuild itself. This is the third step on figure 4.2. The Roussillon plain in the Pyrénées-Orientales developed quickly to become the vegetable garden of France, providing agricultural produce well in advance of the rest of the country. This national 'dependence' on agricultural products from the Pyrénées-Orientales region was highlighted when the 'great flood of October 1940' (Pradé 1941) temporarily caused a halt in production, driving up prices. This incident motivated the Departmental Council to revive a project that had been discussed decades earlier – the Vinça dam.

The need for the dam was socially constructed by the juxtaposition of historical demand for water with 'natural' supply, so it appeared there was a certain scarcity. The suggestion of the need for the dam for the benefit of agriculture strongly reflects the political stance of local politicians. The dam was therefore assigned to flood management (in the spring in particular), and to support irrigated agriculture during the summer.

The fourth point (figure 4.2) describes the second half of the twentieth century, which was a marked period of transformation in irrigated agriculture in the department. The construction of the dam represented strong political state support for the agricultural sector, which experienced a golden age in the post-war period. The dam was an obvious solution for those who sought to secure and expand production. However, as we now know, 1979 was at the tail end of the supply-side paradigm, and a new demand-management paradigm of water management began to take form in many developed countries in the mid-1980s (Delay and Linton 2016; Linton 2010). As a result of diminishing returns of water-supply investments, combined with growing concerns for protecting aquatic ecosystems, the French State gradually made the transition from efforts to augment water supplies to efforts to manage (reduce) water demand – by increasing water productivity – in agriculture and other sectors, often by applying economic instruments to water management.

The filling of the Vinça reservoir in 1979 helped to reduce uncertainty about water supplies. Before its construction 'in the month of July, when water levels were at their lowest, there was not a drop of water' (R. Majoral 2016, personal communication), whereas today water flows in the Têt all year round and 'the younger generations have never experienced shortages' (E. Maillol 2015, personal communication). The most significant change, however, came with the installation of pumps and areas of pressurized irrigation, which allowed irrigation by sprinklers and drop-by-drop systems. To function correctly, the pumps needed to be fed continuously with water, which the dam guaranteed. The installation of pressurized and drip irrigation radically changed the spatial configuration of the irrigation network. It went from a branched structure, intrinsically reflecting upstream–downstream dependence, to a star-shaped network, which eliminated the informal interdependence of – and interaction between – neighbours and harming social structuration (figure 4.3).

Today, the fifth point (figure 4.2), pressurized irrigation and the installation of drop-by-drop systems are widely acclaimed, by virtue of their efficiency, by the Food and Agriculture Organization, the European Union and on a national level by the French *Agence de l'Eau* (Water Agency).[7] In this case, efficiency is understood as the proportion of water withdrawn that is used by plants. The notion of the social efficiency of irrigation is absent. Indeed, the main argument put forward is that of satisfying the needs of the plant without 'waste', in an attempt to modernize agriculture (Lopez-Gunn et al. 2012). However, as E. Maillol (2016, personal communication) underlined, 'The [pumping] station was not to save water; our objective was to make life easier'. Along with the pumping stations, irrigation by atomization allows farmers to reduce their workload and at the same time lessens the upstream–downstream tensions that can arise in times of water shortage (Riaux 2007). However, this change in practice has also had secondary effects, 'Pressurised irrigation eased some tensions, it's true; we no longer depend on our neighbours, but we don't see each other any more either' (E. Maillol 2016).

TORSO for Laponia

Laponia was recognized by UNESCO in 1996, so it was from this date that we applied the TORSO framework to study the evolution of the attitudes in the management structures. However, the national parks that make up Laponia already existed before 1996, so they were actually relatively old in comparison to others in Europe. They were managed by the Swedish Environmental Protection Agency

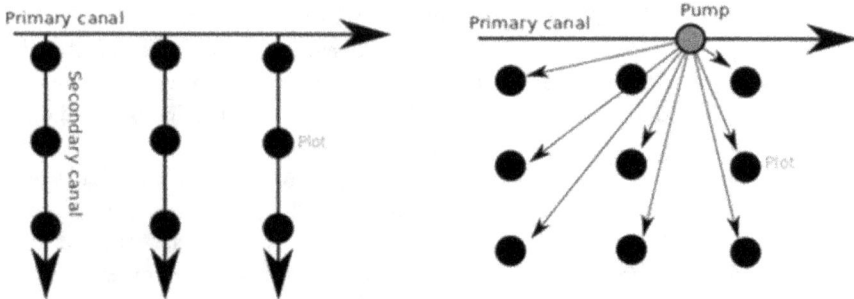

Figure 4.3 Two Patterns of Social Relations Structured by Iirrigation. On the left, the traditional gravity-based system; on the right, a pressurized system.

(SEPA), or *Naturvårdsverket*, which is a state agency. In 1996, the structural organization of the national parks remained the same as it had been before.

There are three steps on the TORSO diagram for Laponia (figure 4.4). The first is for the year 1996 (point 1 on the diagram – figure 4.4) when UNESCO decided that the Laponia area had unique natural and cultural values. However, although UNESCO made the decision to make Laponia a World Heritage Site, it was the Swedish state that had, much earlier, created the borders of the individual national parks. At that time, the indigenous people living there – the Sami – like others in similar cases around the world (Blanc 2015; Brockington et al. 2008) had never been consulted regarding the protection of their land, and most of them did not even know that they were living in a protected area (Green 2009). It is interesting that there was so much confusion in this pivotal year. The SEPA added new rules, the Sami no longer know what they were, or were not, allowed to do, and this was the real beginning of colonialism in the management of previously marginalized spaces (G. Kuhmunen 2017, personal communication). The structure managing the

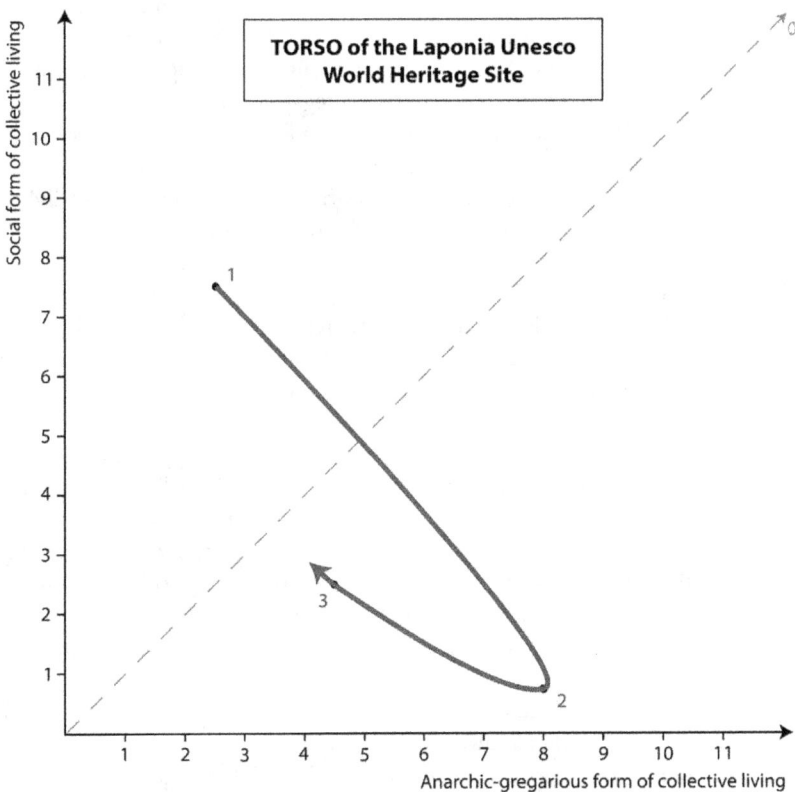

Realisation: S Maraud, E Delay, 2017

Figure 4.4   TORSO for the Laponia UNESCO World Heritage Site.

protected area was very closely related to the state and with the national environmental agency;[8] a very centralized management system with great authority.

Most of the criteria outlined for Laponia prove that variables are very close to the social form. This is why point 1 is located at the top left of the diagram. The idea at that time was to make Laponia a tourist area and that without consulting the local people, who were practising reindeer herding on the land. However, UNESCO encouraged the state to involve the local Sami people in the management and said that without this the World Heritage Site label would be withdrawn. Around the same time, academic works on indigenous issues became more and more prevalent within anti-colonialism movements.

Between the first and the second point, a huge amount of work was done by the local Sami people in order to define what they could do with, and how they could operate within, Laponia (G. Kuhmunen 2017). A long negotiation process took place between the state, the County and the local Sami reindeer herders. Every Sami collective request was automatically refused by the state, which only wanted to employ some Sami individuals, not to question the Swedish structure of the protected area (G Kuhmunen 2017). After years of negotiation, funding was allocated to a group of local Sami to enable them to work on what they could do in Laponia. What happened next was that this reflection, about what the genuine Sami values in the area were, and how elders would have regarded this management project, began a decolonization process. It is quite common for a serious dispute such as this to encourage or stimulate wider thinking about what domination entails, where it comes from and how to counter it. One important Sami value came to light: the *Searvelatnja*. This traditional philosophy is a 'reciprocal learning process' (Balto and Kuhmunen 2014; Sara 2004). The idea is that everyone has a place in the discussion, and knowledge should come from everyone. This represents a large shift in management practice; the state's vision of the land was replaced with a shared vision of the land. The concept of *learning* is essential in this process because it affirms that each stakeholder has his or her own expertise in some areas regarding the land, but also a great lack of knowledge in others. In this case, one has to listen to other visions, practices and attitudes regarding the land in order make appropriate decisions. The territory then becomes a combination of equal territorialities, which makes for suitable and efficient organization. Saul Newman (2001) sees the resistance to domination as the ethical politics of postanarchism. The *Searvelatnja* can easily be compared to this idea. Even though the Sami are not known to have an anarchist background, we can see similarities between anarchism and the ethic of the Sami learning and sharing to understand a place and the practices that take place there (Balto and Kuhmunen, 2014). The mutual aid concept of Kropotkin (1902) also has a lot in common with this philosophy. There, the main principle is to resist against an imposed way of thinking about the land. This ethic of resistance aims to combat the socio-ecological product dictated by the state in order to regain local control of the land and its natural resources (Mullenite 2016).

Over a period of about ten years, as we can see on the second step of the diagram (point 2 – figure 4.4), Laponia underwent a great shift, known as the Laponia

Process. All of the dynamism and reflection about structure and institutional ideology changed the variables considerably, and tended towards the anarchic-gregarious way of collective living. In 2011, there was a need for consensus in making decisions within the new Laponia board: the decentralization of management, the revision of accessibility for local people, the reduction of state authority and of course the introduction of a decision-making system much closer to the associative profile made Laponia an essential decolonizing tool for Sápmi (G. Kuhmunen 2017; Green 2009). Now, the SEPA, the County of Norrbotten, the municipalities of Jokkmokk and Gällivare and the nine Sami reindeer herding communities involved in Laponia are all part of the management of the World Heritage Site. For this reason, this first Sami victory on the Swedish side of Sápmi has become a model to follow, in Sápmi and in the rest of the world, for people in order to fight colonialist states and to have a say in the vision and the practice of nature.

Finally, the last step (point 3 – figure x) is a throwback to the social form tendency. Indeed, six years after the implementation of the Laponia Process, the dynamism decreased. While the Laponia Process was a local decolonizing revolution in 2011, it is now (as o2018) an institution for the practical management of a protected area. Nowadays, the decolonization has to be re-evaluated for several reasons. Many of the anarchic-gregarious attitudes from the second stage did not evolve further, and thus the old order gradually returned. Charles MacDonald explains how implementing order is easier than maintaining equality. The fact that Laponia is now focusing on the practical issues of the area means that autocriticism no longer plays an important enough role. Self-reflexivity about the path of the structure has decreased since 2011, enabling the return of more conservative attitudes to management, which in turn tends to bring collective life closer to a social form.

The accessibility of the structure for the people has not changed, and while in 2011 the involvement of the Sami reindeer herders was a very inclusive procedure, now the question of the legitimacy relating to other local people should be raised. Furthermore, as regards consensus in the decision-making process, new problems appeared after 2011. One of them is the Gállok mine project. A British mining company[9] is in fact asking for permission to exploit an iron deposit 30 km from the Laponia boundary. The infrastructure that the mine will need, as well as the quarry itself, will obviously have an impact on Laponia, or at least on the reindeer in the protected Laponia area.[10] On this issue, Laponia has no say, but the state does. This means, in essence, that although in Laponia there is no hierarchy between the state and the other stakeholders, once the boundaries are crossed, the state returns to being the most powerful decision-maker regarding the management of the land. Thus, if we change the analysis scale, the hierarchy is still the same on strategically important lands (Maraud 2017; Maraud and Desbiens 2017). To conclude, the whole idea of the Laponia Process as a decolonizing tool was that it should be a dynamic expanding process. Yet, for the moment, according to a member of the Laponia board (Dan Ojantva 2017 – personal communication), there is no chance of extending Laponia methods to anywhere else, because Laponia was in an unusually fortunate situation; it had UNESCO support and was in state ownership. Also, the state does not prioritize this type of action, which means that Laponia is very ostracized in Sweden (A. L. Blind 2017, president of the Sami Council, personal communication).

TO REAPPROPRIATE THE TERRITORY

As shown, various criteria can favour the implementation of societal organization profiles. However, this does not mean that the presence of an anarchic-gregarious variable in the criteria automatically implies an anarchic-gregarious collective way of living. Instead, this variable signifies that the criteria move towards this type of collective living and that the implementation of this kind of organization might be facilitated, and vice versa.

The Need of Rational Territoriality

What we call *rational territoriality* is located in the bottom right of the diagram. It means that all the variables of the habitat are known and taken into account in the attitudes regarding the management of the land. This combination of all the territorialities allows a relevant construction of a new territory. Like Doreen Massey (2005), we do think that a space is an accumulation of heterogeneous narratives. However, if the state has the monopoly of the narrative, this implies one territoriality. A rational territoriality means that the narrative monopoly belonging to the state ends in order to make more room to other types of narrative. It would complete the vision of the land and then would make a new form of territoriality. For a rational territoriality, we also have to think about what a territory is, in an exhaustive way of thinking. Ince (2012: 1647) identifies one conception of territory as 'a spatial phenomenon imbued with unequal power dynamics through an (often imposed) territorial imagination of authenticity and belonging'. The territory is about what and who governs it – it is about Foucault's governmentality. This is the recognition of institutions, identities, practices and visions to make a governance model, and also to exclude others. Therefore, territory and the state are not neutral. They both impose an adherence to a social order, and they are both responsible for dissensus (Bourdieu 1990 in 2012). It is thus fundamental for us to reappropriate these notions. The construction of a rational territoriality means both to abolish and to create (Bookchin 1971). As Bookchin demonstrates, with environmental crises, there is no solution without changes in domination and social structures. To rationalize the occupation of the land is the mission of geography (Pelletier 2013), and again it has to be reappropriated so as to not let capitalistic structures to monopolize the debate. To leave the question of resources management to the state leads to the continuation of a system subject to capitalism, whose bases are social organization and domination, with bureaucracy and administration (Pelletier 2013). None of them are neutral; they are instead part of an ideological and imperialist scheme.

The establishment – or not – of authority is essential in the managing attitudes of a territory. If there is one, it imposes an asymmetrical power *via* social relationships and institutions that maintain inequalities (McLaughlin 2007; Ince 2012). The state is one of these institutions that skew the rational territoriality.

Conservative Magnet

TORSO highlights what could be a rational territoriality but, more concretely, the tendency of the trajectory. What is interesting is that both cases tend to go backward in the societal organization, as if there was a *conservative magnet*. By conservatism, we mean behaviour that refuses to change, that needs institutional, ordered control and that rejects diversity and alternative values (De Lagasnerie 2012). By analysing the management of the water in the South of France and of the protected area in Laponia, we found out that some main characteristics of the social form of collective living are back in the structures. A return to bureaucracy, hierarchy, specialization and authority seems to reoccur as soon as resistance to maintaining equality loses its dynamism. This conservative magnet does seem to exist in many trajectories; it is the ability of capitalism to absorb spaces of resistance.

We found that this conservative magnet was also active in our own method. Indeed, the perpetual reference to the dominant society in our definition of what is 'an alternative' (Lacroix 1981) might be seen as an issue in our work. On the contrary, we do think it quite interesting to see how powerful social learning can be. To see the territory as the state and the institutions as a bureaucratic administration show how socialization – by schools, the media, the family and the social arena – influences our own perspective of a structural alternative. The institutional and hierarchic revival in our two examples shows that as soon as the dynamism and the effort to maintain an anarchic-gregarious tendency decrease, the conservative magnet pulls the society back towards a social form of organization. But it also means that there is a need to decolonize education by reappropriating notions and debates. An institution is seen as a bureaucratic entity frozen in the past and depending on the state, but it could be a discussion arena to recreate the territory through the people living on the land (Ince 2012).

Being conscious of our conservative logic in our work is crucial, there is thus a need to include self-critic in the analysis. Integrating this element is part of our work; the researcher is not neutral and not external to any processes he or she denounces (Bourdieu 2001, 2003).

Avoiding Fatality Theories

Although the results in our two cases demonstrate a return to a social-hierarchic way of collective living, the aim of using TORSO is not to prove that conservative attraction is an automatic societal trend. Since our work is based on contextualized studies, our approach differs from theories such as the 'Iron law of oligarchy' (Michels 1911) or the 'Institutional isomorphism' (DiMaggio and Powell 1983).

Indeed, even though these two theories are very interesting, it seems unsafe to make them universal (Bourdieu 1997). In summary, the Iron law of oligarchy shows that every organization will create domination relationships and the emergence of an elite. Institutional isomorphism theorizes the homogenization of organizations caused by three mechanisms: a coercive isomorphism, a mimetic isomorphism and a normative isomorphism (DiMaggio and Powell 1983). These three pressure instruments that make the institutions of a same scope similar. However, we think that while fatalist laws regarding alternatives are interesting from a critical perspective they also serve as a mean to denigrate other already existing forms of living. Anarchic-gregarious attitudes are a way to disprove these laws. On the other hand, it is always important to keep a critical approach regarding these alternatives to avoid oligarchic processes and isomorphism within the collective living implementation. This is the role of TORSO to identify and to understand the elements that make the trajectory of an organization tend towards another. TORSO is not about absolute theory, but is about decoding and emphasizing attitudes in order to amplify or counteract them. As for social science, it is about understanding the reasons of human behaviour (Bourdieu 1997).

However, in addition to the fact that TORSO is a social scientific tool, it is also an anarchist one. Anarchism is a dynamic for liberty. The territory is not an entity frozen in the past; it is perpetually in construction, with the redefinition of values and territoriality (Springer 2014; Ince 2012). Currently, it is characterized by a hierarchy of constitutive values, but there is a need to end these hierarchies in order to have each value proper to local stakeholders in the architecture of the territory (Ince 2012; Valentine 2007; Marston 2000). The state, as one of the main incarnations[11] of authority, is responsible for the unbalanced representation of values in the structures and the institutions of management. TORSO focuses on abuse, perpetrated by the state, which prevents the dynamic and logical evolution of the territory, embodied by its inhabitants. Thus TORSO is a tool for anarchist key concepts: liberty and autonomy.

Potential Participative Tool

To conclude, one main criticism against the use of frameworks is that that they can be interpreted as a wish to put a halt to conflicts about places (De Lagasnerie 2012). In use, the TORSO framework can be developed as a participative tool. Using a companion-modelling approach (Etienne et al. 2013), the TORSO framework can be rebuilt in a participative way. Involvement of stakeholders from the very beginning of TORSO will introduce a way to understand their points of view, of social organization in a context of natural resources management. TORSO is intended to enable reflection about the attitudes of emancipation in order to make the trajectory analysis understandable, and in a pedagogic manner. It is true that an academic framework can be seen as a means of domination that tries to categorize realities. This is why we tried to design this framework as something that can be reappropriated by people. As a next step, it would be interesting to think through the criteria and the attitudes with the population concerned, in order to compare results and to understand their own vision of a phenomenon. TORSO has great participative potential to help achieve a coherent approach to the contemplation and constitution of new territories.

By being critical of emancipatory trajectories, we show failures and instrumentalizations but also successes. There is a need to rehumanize societies (Gerber 2015 in Bookchin) and to politicize stakeholders again, particularly their actions. Bookchin (1971) said change necessarily implies an institutional shift; TORSO illustrates this shift by redefining the attitudes and also the many obstacles faced in this complex process.

NOTES

1. Particularly for the cultivation of fruit and forage, which have been the outstanding products of the region
2. Since the Second World War, there has been no census regarding the number of Sami people, only estimations. After all the racist policies implemented by the Nordic States, it has been decided not to survey the Sami in order to avoid this taboo topic.
3. This distinction is very important because the reindeer herders represent only 10 per cent of the Sami in Sweden (Roturier and Roué 2015).
4. This chapter will focus on the Sami and the irrigants, but in future works, we are trying to apply TORSO also with the Cree in Northern Canada and the Peuls in Senegal.
5. As the people living in anarchy but without political doctrine.
6. The fact that the institutions are specialized accentuates the social form of living because it makes a problematic even more complex in its handling.
7. For example, on the water agency website, drop irrigation is presented as a solution for water efficiency. https://www.eaurmc.fr/jcms/vmr_36233/fr/mieux-partager-l-eau-mieux-mai triser-les-besoins consulted 08/04/2018.
8. Which is the SEPA.
9. For example, http://beowulfmining.com/projects/sweden/kallak/, last visit in November 2017
10. With the transhumance, the reindeers go from winter land to summer land, and Gállok is located on the grazing land they cross twice a year.
11. It also could concern international firms for instance.

REFERENCES

Balto, A. M., & Kuhmunen, G. (2014). *Máhttáhit ieaamet ja earáid – Máhttáhit : omskola dem och oss – Máhttáhit : re-educate them and us! Sami self- determination, nation-building and leadership.* Karasjok, Norway: ČálliidLágádus.

Blanc, G. (2015). *Une histoire environnementale de la nation: regards croisés sur les parcs nationaux du Canada, d'Éthiopie et de France.*

Bookchin, M. (1977). *Post-Scarcity Anarchism* (2nd ed.). Black Rose Books, No. 071. Montréal: Black Rose Books.

————. (1990). *Remaking Society: Pathways to a Green Future*. F Second Printing Used edition. Boston, MA: South End Press.

————. (2004). *Post-Scarcity Anarchism* (new ed.). Edinburgh; Oakland, CA: AK Press.

Bourdieu, P. (2001). *Science de la science et réflexivité*. Paris: Raisons d'agir.

————. (2003a). «L'objectivation participante». *Actes de la Recherche en Sciences Sociales* 150(1): 4358. https://doi.org/10.3406/arss.2003.2770.

————. (2003b). *Méditations pascaliennes*. Édition revue et corrigée. Paris: Seuil.

————. (2012). *Sur l'État. Cours au Collège de France*. Paris: Le Seuil.

Brockington, D., Duffy, R., & Igoe, J. (2008). *Nature Unbound: Conservation, Capitalism and the Future of Protected Areas*. London; Sterling, VA: Routledge.

De Greef, G. (1895). *Le transformisme social : essai sur le progrès et le regrès des sociétés/par Guillaume de Greef,...* Paris: F. Alcan. http://gallica.bnf.fr/ark:/12148/bpt6k5512606n.

Delay, E., & Linton, J. (2016). The Vinca Dam and the Withering of Canal Associations in the Têt Basin of the Eastern French Pyrenees. In *Water and Social Relations: Wittfogel's Legacy and Hydrosocial Futures*. University of Milano-Bicocca.

DiMaggio, P. J., & Powell, W. W. (1983). The Iron Cage Revisited: Institutional Isomorphism and Collective Rationality in Organizational Fields. *American Sociological Review* 48(2): 147160. https://doi.org/10.2307/2095101.

Escobar, A. (1996). Construction Nature: Elements for a Post-Structuralist Political Ecology. *Futures* 28(4): 325343. https://doi.org/10.1016/0016-3287(96)00011-0.

Etienne, M. & Collectif. 2013. *Companion Modeling: A Participatory Approach to Support Sustainable Development*. Springer Science & Business Media.

Gibson, T., & Sillander,K. (Ed.). (2011). *Anarchic Solidarity: Autonomy, Equality, and Fellowship in Southeast Asia*. New Haven, CT: Yale University Southeast Asia Studies.

Green, C. (2009). *Managing Laponia: A World Heritage as Arena for Sami Ethno-Politics in Sweden*. Uppsala, Sweden: Uppsala Universitet.

Green, K. E. (2016). A Political Ecology of Scaling: Struggles Over Power, Land and Authority. *Geoforum* 74(août): 8897. https://doi.org/10.1016/j.geoforum.2016.05.007.

Ince, A. (2012). In the Shell of the Old: Anarchist Geographies of Territorialisation. *Antipode* 44(5): 16451666. https://doi.org/10.1111/j.1467-8330.2012.01029.x.

Ingold, A. (2009). Les sociétés d'irrigation : bien commun et action collective. *Entreprises et histoire*, n° 50 (février): 1935.

Jaubert de Passa, F. (1821). *Mémoire sur les cours d'eau et les canaux d'arrosage des Pyrénées-Orientales*. Memoires d'agriculture d'économie rurale et domestique. Huzard.

Kropotkine, P. (1892). *La Conquête du Pain*. Dialectics.

————. (1901). *La science moderne et l'anarchie*. Antony: Tops/H.Trinquier.

————. (1902). *L'entraide : Un facteur de l'évolution*. Montréal: Ecosociété.

————. (1921). *L'Éthique*. STOCK. Paris: Stock.

Lacroix, B. (2006). *L'utopie communautaire: Histoire sociale d'une révolte* (2e edition). Paris: Presses Universitaires de France – PUF.

Ladki, M., Guérin-Schneider, L., Garin, P., & Baudequin, D. (2012). Des canaux d'irrigation aux canaux de distribution d'eau brute ? In *De l'eau agricole à l'eau environnementale. Résistance et adaptation aux nouveaux enjeux de partage de l'eau en Méditerranée*, par Chantal Aspe, 1934. Update Sciences & technologies. Edition Quae. http://www.cairn.info/resume.php?ID_ARTICLE=QUAE_ASPE_2012_01_0019.

Lagasnerie, G. (2012). *La derniere leçon de Michel Foucault: Sur le néolibéralisme, la théorie et la politique.* Paris: Fayard.

Leitner, H., & Miller, B. (2007). Scale and the Limitations of Ontological Debate: A Commentary on Marston, Jones and Woodward. *Transactions of the Institute of British Geographers* 32(1): 116–125.

Linton, J. (2010). *What is Water?: The History of a Modern Abstraction.* Nature|History|Society. Vancouver, Canada: UBC Press.

Lopez-Gunn, E., Zorrilla, P., Prieto, F., & Llamas, M. R. (2012). Lost in Translation? Water Efficiency in Spanish Agriculture. *Agricultural Water Management, Irrigation Efficiency and Productivity: Scales, Systems and Science* 108(mai): 8395. https://doi.org/10.1016/j.agwat.2012.01.005.

Macdonald, C. (2011). Anarchs and Social Guys. *Society* 48(6): 489494. https://doi.org/10.1007/s12115-011-9483-y.

———. (2013). The Filipino as Libertarian: Contemporary Implications of Anarchism. *Philippine Studies: Historical and Ethnographic Viewpoints* 61(4): 413436. https://doi.org/10.1353/phs.2013.0025.

———. (2014). Anthropologie de l'anarchie et anarchisme. *Conférence au Centre International de recherches sur l'anarchisme.* Marseille.

———. (2016). Structures des groupes humains, Structural Properties of Human Organizations. *L'Homme*, n° 217(février): 720.

Maraud, S. (2017). L'intégration des Autochtones : vers des territoires post-coloniaux ? La Nature et la culture comme outils décoloniaux chez les Samis et les Cris. *INDITER-Territoires et territorialités autochtones.* Lille. France.

Maraud, S., & Desbiens, C. (2017). Eeyou Istchee – Baie James, vers un capital environnemental mixte ? *Norois. Environnement, aménagement, société*, n° 243 (novembre): 7188. https://doi.org/10.4000/norois.6095.

Marston, S. A. (2000). The Social Construction of Scale. *Progress in Human Geography* 24(2): 219242. https://doi.org/10.1191/030913200674086272.

Martinez-Alier, J. (2002). *The Environmentalism of the Poor: A Study of Ecological Conflicts and Valuation.* Edward Elgar Publishing.

Matteuzzi, A. (2012). *Les Facteurs de L'Evolution Des Peuples: Ou, L'Influence Du Milieu Physique Et Tellurique Et De L'Heredite Des Caracteres Acquis Dans L'Evolution Et L.* Nabu Press.

McLaughlin, P. (2007). *Anarchism and Authority: A Philosophical Introduction to Classical Anarchism* (1st ed.). Aldershot, Hants, England; Burlington, VT, USA: Routledge.

Metchnikoff, L. (1886). Revolution and Evolution. *Contemporary Review* 50: 412–437.

———. (1889). *La Civilisation et les grands fleuves historiques.* Paris: Hachette.

Michels, R., & Lipset, S. M. (2016). *Political Parties: A Sociological Study of the Oligarchial Tendencies of Modern Democracy.* Traduit par Eden Paul. Martino Fine Books.

Mullenite J. (2016). Resilience, Political Ecology, and Power: Convergences, Divergences, and the Potential for a Postanarchist Geographical Imagination. *Geography Compass* 10(9): 378388. https://doi.org/10.1111/gec3.12279.

Neumann, R. P. (2009). Political Ecology: Theorizing Scale. *Progress in Human Geography* 33(3): 398–406.

Newman, S. (2001). *From Bakunin to Lacan: Anti-Authoritarianism and the Dislocation of Power.* Lexington Books.

Pelletier, P. (2013). *Géographie & Anarchie, Reclus, Metchnikoff, Kropotkine.* https://halshs.archives-ouvertes.fr/halshs-01251723/.

Proudhon, P. J. (1840). *Qu'est-ce que la propriété ? ou Recherches sur le principe du droit et du gouvernement, par P.-J. Proudhon... Premier mémoire.* Paris: J.-F. Brocard. http://gallica.bnf .fr/ark:/12148/btv1b8626552d.

Riaux, J. (2007). "La reproduction des eaux par les arrosages", historique et actualité d'une théorie. *Conserveries mémorielles. Revue transdisciplinaire,* n° #2 (janvier). https://cm.revues .org/171.

Rist, G. (2007). *Le développement: Histoire d'une croyance occidentale* (3e édition revue et augmentée). Paris: Les Presses de Sciences Po.

Roturier, S., & t Roué, M. (2015). Le Pâturage, C'est toute une science ! *Techniques & Culture,* n° 63: 92109.

Sara, M. N. (2004). Samisk kunnskap I undervisning or læremidler. Hirvonen, Vuokko (red) Samisk skole I teori og praksis. Karasjok. *ČálliidLágádus* 114–130.

Springer, S. (2014). Human Geography Without Hierarchy. *Progress in Human Geography* 38(3): 402149. https://doi.org/10.1177/0309132513508208.

———. (2016). *The Anarchist Roots of Geography: Toward Spatial Emancipation* (1re éd.). Minneapolis: University of Minnesota Press.

Valentine, G. (2007). Theorizing and Researching Intersectionality: A Challenge for Feminist Geography. *The Professional Geographer* 59(1): 1021. https://doi.org/10.1111/j.1467-9272 .2007.00587.x.

5

When the Wolf Guards the Sheep

The Industrial Machine through Green Extractivism in Germany and Mexico

Alexander Dunlap and Andrea Brock

Rick: I then introduced that life to the wonders of electricity! Which they now generate on a global scale and you know some of it goes to power my engine and charge my
phone and stuff.
Morty: You have a whole planet sitting around mak'n power for you? That's slavery!
Rick: It's society . . . They work for each other, Morty. They pay each other. They buy
houses. They get married and make children that replace them when they get too old to
make power.
Morty: That just sounds like slavery with extra steps.

— *Rick and Morty*, 'The Ricks Must Be Crazy'

The time has come to recognize industrialism, modernization, development and
'progress' – as it is euphemistically known – as the greatest threats faced by the
Earth and its inhabitants. The interrelated and self-propelling ecological, climate
and economic crises – the outgrowth of evolving processes of patriarchy, slavery,
white supremacy, ecocide and genocide, and a prerequisite for state formation – are
manifestations of this threat (Davis & Zannis 1973; Perlman 1985; Moses 2008;
Gelderloos 2017; Scott 2017; Öcalan 2013). While industrial development has not
been acknowledged as 'the problem', governments now recognize mass extinction
and climate catastrophe – in a narrower sense – as significant threats to the existing
political economic order. They recognize, at least discursively, the need to phase
out fossil fuels to satisfy the never-ending thirst for energy to power the capitalist

mega-machine (Mumford 1967/1970) and push – together with many environmental campaigners – for a shift towards 'clean' or 'green' energy sources: renewables.

Taking an anarchist political ecology perspective (Brock 2020b; Springer 2022), this chapter examines the confluence of what mainstream environmental (justice) activists might call the 'problem' – fossil fuel extraction, particularly coal-mining – and the 'solution' – renewable energy such as wind energy. We put forward the notion of the 'renewable energy–extraction nexus' to critique the continued reliance on extractivism and ecosystem exploitation that are fundamental to the renewable energy infrastructure supply chains and to highlight the parallels between renewables and extractive industries via two case studies: coal-mining in the German Rhineland and wind energy development on the Mexican Isthmus. Both, we show, are (and continue to be) integral to a new 'green economy' that reinforces statist imaginaries and industrial ideologies that attempt to obscure, invisibilize, and consequently renew socio-ecological destruction.

In the present conversation around climate change mitigation, the coal and wind energy industries are positioned as diametrically opposed and often compete over state subsidies and market shares. Policymakers, corporate decision-makers, researchers, policy advisors and environmental NGOs tend to share the enthusiastic embracement of renewable energy technologies to break with fossil fuel dependence and unsustainable energy production.[1] The necessity of a 'move' to 'clean' renewables, which would magically replace 'dirty' fossil fuels, is taken for granted. The messy political history of so-called energy transition remains overlooked (Bonneuil 2016; Smil 2016a). Even in environmental justice circles, critiques of renewables are often met with fierce opposition.

On further analysis, however, we argue, industrial-scale and corporate-controlled renewables and fossil fuels are accomplices in the struggle to control, usurp and transform the vitality of the natural environment. Coal-mining and wind energy are constitutive of the trajectory of ecocide and a multiplicity of slaveries emblematic of modernity. We draw on Bram Büscher and Veronica Davidov's 'ecotourism–extraction nexus' that demonstrates how resource extraction and ecotourism are actually co-constructed, share similar logics and retain multiple forms of collaboration (Büscher & Davidov 2013; Brock 2020c). The renewable energy–extraction nexus extends this concept to renewable energy.

The rise of renewables, we argue – as part of climate change mitigation strategies – is embedded in the hegemonic logic of green capitalism. Ideas of 'sustainable development', the 'green economy', 'ecosystem services', 'smart agriculture' and 'resilience' have all been positioned to enable the continuation of capitalist development under the name of climate change mitigation, conservation and/or adaptation (Dunlap & Fairhead 2014; Hunsberger et al. 2017). The green economy not only attempts to reconcile ecosystem health and capitalist development, but it also offers new natural resource valuations that create new markets and opportunities for expanding economic growth. This entails integrating previously excluded non-human natures or, in economic jargon, 'market externalities' into economic logics and accounting practices. The green economy is the economy that now recognizes, includes and consequently further intensifies the exploitation of 'nature', enmeshing further natural resources into the machinations of economic and financial structures (Fairhead et al. 2012; Corson et al. 2013; Dunlap & Fairhead 2014; Dunlap 2019). A notable machination is the discursive transformation of nature into 'ecosystem services' or 'natural capital', which necessitates the further spread and entrenchment of enclosures, greater ecosystem surveillance and the cataloguing and discursive fabrication of nature as a commodity service to become commensurable and tradable within financial markets (Lohmann 2008; Sullivan 2010, 2013a, 2017; Dunlap 2019). This transformation of flora and fauna into carbon, biodiversity and other so-called environmental commodities allows the enactment of 'offsetting' logic, which assert that ecological destruction can be compensated with payments towards emission reductions or environmental-engineering initiatives to create ecological improvements in new or existing environmental sites (Sullivan 2010; Böhm & Dabhi 2009; Brock 2015, 2020c). The creation of 'new natures' through restoration activities is often accompanied by a large-scale land dispossession to facilitate 'No Net Loss', 'land degradation neutrality' or 'carbon neutrality', and further forms of 'accumulation by restoration' (Huff & Brock 2017; Brock 2020c). In short, offsetting is a crucial mechanism that claims to reconcile capitalist development with nature 'conservation', which has become increasingly popular with extractive (and other) industries.

Currently, industrial-scale renewable energy generation relies on – and is co-constructed by – continued hydrocarbon and mineral extraction processes and conventional energy infrastructures. Rather than breaking with the logics, power relations and processes of fossil fuels, they deepen the existing political economy of energy, processes of dispossession, destructive social and ecological relations, and accumulation. Providing two case studies from different extractive industries, cultural contexts and countries, we place coal-mining in Germany and wind energy development in Mexico side by side to examine key features of the renewable energy–extraction nexus emerging across sites. The studies are built on extensive field research, with field visits and contacts with people in these areas. We draw on participant observation, public events, informal and semi-structured interviews[2] in addition to secondary research material including books, newspaper articles, promotional materials and blogs.

We begin our chapter by first offering some principles from green anarchy to develop important values for an anarchist political ecology critique of the renewable energy–extraction nexus, illuminating neglected issues that highlight the colonial nature of the industrial system responsible for the present state of ecological and climate crisis. After highlighting the normalizing and self-reinforcing nature of industrial systems, we turn to examine RWE's mining operations in the German Rhineland. We discuss RWE's Hambach mine, the world's largest opencast lignite coal mine that – while strongly resisted – is slowly destroying large parts of the ancient *Hambacher Forest*. This destruction is justified by RWE's deployment of green economic technologies of governance including nature recultivation and offsetting initiatives (Brock 2020c) and legitimized by their corporate social technologies that attempt to marginalize and pacify militant resistance in the area (Brock & Dunlap 2018). After delving into RWE's attempts at 'sustainable mining', we then turn to the largest wind energy (factory) development in Latin America, the Isthmus of Tehuantepec region of Oaxaca, Mexico – known locally as the *Istmo*. Regarded as a climate change mitigation strategy, wind energy in the *Istmo* has similar impacts to traditional extractive industries not only in the ways developers acquire land but in relation to the violence dispensed against local indigenous groups contesting the construction of these projects. The next section will compare and discuss the similarities, differences and relationships between coal and wind energy extraction. Here, we coin the 'renewable energy–extraction nexus' to describe how conventional and renewable energy systems are dependent on each other, collaborate, and together expand and intensify industrial development and socio-ecological degradation in a rush to grab all the vital energies of the Earth. We conclude by arguing that the green economy is renewing destruction not only by 'greening' – thus legitimizing – inherently unsustainable industrial activities but by expanding such activities and relationships at the cost of social and ecological diversity and health. Value is extracted from the process of 'greening' itself, while industrial systems continue to exercise 'war by ecological crisis'.

NEITHER MARKET NOR STATE: SHEEP AGAINST INDUSTRIAL PROGRESS

Michel Foucault's genealogy of government locates the root of government in the Christian shepherd-flock analogy: god is the shepherd of 'men', and the shepherd (with 'his' connection to god) is the governor of the flock. In his reading, government becomes the shepherd and the population becomes the flock (Foucault 2007). The green economy, then, is akin to letting a pack of wolves guard the sheep or, more accurately, letting governments and corporations organize ways to manage and 'repair' ecologically and socially disastrous life forms that they themselves have organized for so long. What is the goal or the endgame of 'society', the 'state' the 'government'? Instructive is verse 1:28 in *The Book of Genesis*: 'Be fruitful and multiply, and fill the earth and subdue it; and have dominion over the fish of the sea and over the birds of the air and over every living thing that moves upon the earth.'

Plenty of work has illustrated the violence and hierarchical ordering foundational to the state system, and government as one of its manifestations (Gorz 1980; Ince & Barrera de la Torre 2016; Scott 1998). This violent ordering is integral to the statist system: the 'pervasive, historically contingent organizational logic that valourises and naturalises sovereign, coercive, and hierarchical relationships, within and beyond state spaces' (Ince & Barrera de la Torre 2016). The state system, the capitalist economic system, and the industrial order and ideology that it protects and relies upon are themselves the product of – and reproducers of – colonial mindsets and practices reliant on the exploitation of humans and non-humans alike. Scholars have long identified the continuities and intricate relationship between capitalism, industrialization and slavery – and especially plantation slavery as essential to US American capitalism (Walter 2013). C. L. R. James and Eric Williams first argued for the recognition of the centrality of slavery to capitalism and 'modernity' over eighty years ago. By showing how Atlantic modernity was constructed through engagement with colonial capitalism in the West Indies, James argued that slavery was a product of Renaissance rationality (James 1938), while Williams explored the relationship between colonial development and European industrialism to illustrate the contradictions in modernist rationality (Williams 1944). Indeed, global capitalist development was fundamentally dependent upon colonial appropriation and exploitation, and 'colonial processes are also central to the production of racialized inequalities upon which capitalism is itself structured' (Bhambra 2020, 14; Rodney 1972/2009). Plantations and plantation slavery were key to the development of modern scientific management techniques (Rosenthal 2016) and profits from slave trade and plantations were a financed Britain's nascent industries (Williams 1944), religious institutions, hospitals, railways (Karuka 2019) and more.

Contemporary, or 'new capitalism', according to Sven Beckert, characterized by wage labour and states' unprecedented bureaucratic, infrastructural and military capacities 'had been enabled by the profits, institutions, networks, technologies, and innovations that emerged from slavery, colonialism, and land expropriation' (Beckert 2014a: n.p.; see also Baptiste 2014; Clegg 2015; Beckert 2014b). Capitalism itself, David Graeber famously argued, constitutes a continuation of slavery in a broad sense, as 'any form of labor in which one party is effectively coerced' (Graeber 2006, 68–69). They share the reliance on separation of place of production and social reproduction, the exchange of human powers for money, the requirement of the social death of workers/slaves, the production of 'abstract labor' and the embedding in an 'ideology of freedom' (Graeber 2006, 79).

Governments and their apparatuses of administrative decentralization – based on multiple systems of oppression such as race, gender, class and speciesism – drive political stability, industrial 'progress' and organizational expansion (Dunlap 2014b). These forces seek constant organizational self-affirmation, guarding their existence and expanding their mentality, relationships and purpose across the world. This religious drive manifests itself as economic growth, urbanization and infrastructural development that require constant mining, processing, manufacturing and consumption of natural resources, both human and non-human (Dunlap & Jakobsen 2020; Springer Volume 1).

Majid Rahnema (1997) and Lorenzo Veracini (2014) demonstrate the viral and bacterial qualities of colonization and development that receive little attention or redress, instead provoking cognitive dissonance from those immersed in industrial life – life infected by rhetoric of 'peace', technological enchantment and ideas of 'progress', economic or otherwise. Organizational stability and qualitative and quantitative growth are the modus operandi of modernity and consumer society, which leads to two foundational insights for anarchist political ecology. The first is explained well by Kirkpatrick Sale (1991/1985, 122), summarizing Murray Bookchin:

[S]ocieties that dominate nature also dominate people. Where there is the idea that a massive dam should be built to control a river's flow, there is the idea that people should be enslaved to build it; where there is the belief that a giant metropole may serve itself by despoiling the surrounding countryside and devouring its raw materials, there are castes and hierarchies to ensure that this is accomplished.

Embedded here is Elisée Reclus' realization that *humans are nature* or 'nature becoming self-conscious' (2013/1905, 3) and Mikhail Bakunin's notion that 'every enslavement of men [*sic*] is at the same time a limit on my own freedom' (2005/1871, xi), as these notions are applied to non-human life and megaprojects from ancient civilization to present. Eco-anarchism, as John Clark reminds us, 'is the form of political ecology that situates the political most deeply in Earth history and in the crisis of the Eart' (Clark 2020, 9–14). Yet, capitalist development has instilled the exact opposite idea: the more non-human and human lives are enslaved and consumed, the greater 'freedom' one is meant to obtain. While the fruits of modern life – cars, planes, computers, microwaves – symbolize this new freedom (for those who can afford them), these liberties are intrinsically enmeshed with military conquest, classical and modern slaveries (Fitzpatrick 2018), and ecocide that have become historically justified,[3] erased or made seemingly 'irrelevant' in everyday life (Dunlap 2020).

The 'natural resource base upon which industrial societies stand is constructed in large part through the use and threatened use of armed violence', Liam Downey and colleagues have argued, and it 'quickly becomes apparent that armed violence and the environmental degradation associated with it are intimately woven into the everyday lives of core nation citizens through the purchases they make and the fuels they consume' (Downey et al. 2010, 437–438). Furthermore, Tanya Li writes, 'When the land is needed but labour is not, the most likely outcome is the expulsion of people from the land' (Li 2011, 286), often by the military or other violent forces. This sounds oddly familiar to A. D. Moses's discussion of Jean-Paul Sartre (1968) and the politics and methods of post–Second World War genocide where 'physical annihilation was checked by the need for indigenous labour', as colonial powers' response 'to the inevitable guerrilla resistance was to annihilate part of the population in order to terrorize the rest' (Moses 2002, 24) into submission to a colonial (producer–consumer) paradigm.[4] This connection between widespread political violence and ecological degradation or the 'genocide–ecocide nexus', as Damien Short (2016) calls it, plays a fundamental role in the colonial and, by extension, industrial progress that takes on increasingly complicated, yet progressive forms (Dunlap 2018; Brock 2020b). In sum, the continuation of the present trajectory of industrial and computational development requires increasing methods of strategic violence and ever-more sophisticated forms of participatory slavery that are deeply intertwined with dependency, addiction and systemic path dependency.

This leads to the second point: the recognition that both ancient and industrial civilizations and later forms of state organization, in all of their varieties, are inseparable from colonialism and the colonial model (Dunlap 2014b, 2018; Katsiaficas 2006/1997; Springer 2016). Explaining the complexities and continuities of colonial genocide, Patrick Wolfe reminds us, 'invasion is a structure not an event' (Wolfe 2006, 388). It is from this perspective that we assert that industrial development itself is a system of domination, which domesticates humans and non-humans, assimilates difference and transforms ecosystems to a point of severe degradation or destruction. The evolution of industrial development has necessitated various political modes of governance and politics – autocracies, oligarchies and democracies – that always required some form of slavery or exclusion (Ellul 1964/1954; Landstreicher n.d.; Gelderloos 2013; Güven 2015).

Transcending every type of capitalism – Keynesian, command control, neoliberal, financial and so on – a focus on industrialism (Brock 2020b) allows us to peer into the core of capitalism and its material embodiments – guarding against state and working-class romanticism. It is industrialism itself, and its political and culture industries, that manufacture desire and consent,[5] imbuing human dependency, addiction and normalization of political, economic and industrial structures (Porter & Kakabadse 2006; Alexander 2008; Paoli 2013/2008). The normalization of industrialism in everyday life prevents even critical scholars from acknowledging their implicit statists and industrial subjectivities (Ince & Barrera de la Torre 2016). Such acknowledgement means resituating how we view electricity (Winther 2008), sanitation systems (Dunlap 2017a), roads (Dalakoglou 2012) and other industrial infrastructural amenities. It demands analysing them as systemic techniques of integration and domestication to create and reproduce an intricate, energy-intensive network that justifies, enables and spreads industrial relations and infrastructure (Berman 1983; Cullather 2006, 2013/2010). In short, industrialism constitutes the material practice of conquest, otherwise known as the industrial, social (Dunlap 2014a) or genocide machine (Davis & Zannis 1973), which –despite its negative social and ecological consequences – is becoming rebranded as 'sustainable' and 'green' that the study of political ecology has revealed so well.

The remainder of this chapter explores how the industrial system continues in the face of ecological, climate and economic crisis, or how a cunning wolf can become a shepherd in charge of the flock, by investigating the renewable energy–extraction nexus through coal-mining in Germany and wind park development in Mexico. We focus specifically at the material heart of the industrial system, which is extractivism – the mining of the Earth and harnessing of wind.

GREENING DESTRUCTION: DISCIPLINING AND DOMESTICATING HUMAN AND NON-HUMAN NATURE IN AND AROUND THE HAMBACH COAL MINE

With a size of 85 square kilometres or 8,500 hectares, the Hambach coal mine (figure 5.1) is known to be the largest human-made hole in the world (Der Spiegel 1982). Throughout its lifetime, the open-pit lignite mine that is 'migrating' across the Rhineland has been responsible for the forced displacement of thousands of

Figure 5.1   Hambach Mine. *Source*: Hubert Perschke (2012).

people, the destruction of one of Europe's most ancient and biodiverse forests (the *Hambacher Forest*) and the release of more greenhouse gas emissions than any other industrial project in Europe (Brock & Dunlap 2018). Despite intense resistance against the mine – forest occupations, demonstrations and sabotage, among others (Anonymous 2016) – the social and ecological disaster in and around the mine continues to unfold.[6] The defence of the mine at any cost by police and corporate security forces is embedded in Germany's long tradition of surveillance of resistance, intensifying social control and increasingly visible authoritarian state structures leading to violent responses to any kind of contestation.[7] Most recently, this can be seen in the violent repression of environmental defenders attempting to stop the construction of a new highway in 2020 (Brock 2020a) as well as the well-documented state violence against G20 protesters and their long prison sentences in 2017 (NDR 2017).[8] Both point to the hypocrisy of the image of the German state as both socially and environmentally progressive. The former state-owned electricity provider, coal mine operator, nuclear and renewable energy producer, and self-proclaimed 'energy giant' RWE and its shareholders continue to benefit from political support for ecologically destructive activities like coal-mining and their close ties to the political establishment (Brock & Dunlap 2020).[9]

The Rhineland thus serves as a great example to illustrate how the ecological crisis is discursively acknowledged and subjected to policymaking while extractive interests continue to be protected. The German state reconciles commitments to climate change mitigation by mobilizing 'green' technologies and marked-based mechanisms through promotion of renewables, e-mobility and carbon pricing as solutions while selling new, 'cleaner' coal power stations as contributions to climate protection[10] in an attempt to fragment popular contestation and to ensure public support for mining.

RWE's work to rebrand mining as 'sustainable' is justified by scientific abstractions and calculations that focus on singular aspects of ecosystems – 'carbon' and 'biodiversity' – that neglect numerous issues and qualities associated with interventions into ecosystems (Brock 2020c). Continued extractivism is possible because it lies not only at the heart of industrial production but at the heart of modernist ideology and the state system; involving not only the mining of (fossil) resources but also the capturing of hearts and minds of the population. Green extractivism, or green mining, is thus central to the reconciliation of industrial destruction with social and ecological 'sustainability' in the form of the 'green economy'. In the Rhineland, this occurs through a number of mechanisms: first, the anchoring of RWE as 'good corporate neighbour' and responsible employer with the best interest of local communities and the wider German public at heart. Second, sustainable extractivism is constructed through 'green-washing' of its operations and supply chains to ease concerns around ecological impacts and human rights violations which go hand-in-hand with the spectacularization (Debord 1967) and commodification of the mining experience (Brock 2020c). Third, RWE is able to appear 'progressive' and 'environmentally responsible' through divide-and-conquer strategies to manage resistance against coal-mining; from engagement with conservation organizations to the criminalization of and physical violence against forest defenders.

Positioning RWE as responsible corporate neighbour and indispensable partner for the German public involves substantive investments into a 'Public Relations war' to win the hearts and minds of the population (McQueen 2015), ensuring loyalty to RWE's corporate brand and engineering and buying consent to its projects. Propaganda 'is cheaper than violence, bribery and other possible control techniques' (Lasswell 1934, 524) or, in the words of Paul Virilio, '[b]eating an enemy involves not so much capturing as captivating them' (Virilio 1995, 14). Beyond RWE's Public Relations campaigns (through the advertising and marketing industries and associated consultancies), this involves investments into astroturfing (the set-up of fake citizens' associations), lobbying efforts in schools, sponsorships of community, police and fire service events, sports clubs and school projects, among many others.[11] These efforts to win over the population extend to investments into new recreational infrastructure including a huge cycling and hiking network, cultural activities, museums, exhibitions and financial support for stadiums as well as school projects. At the same time, RWE has worked hard to dismiss concerns about 'irreparable ecological consequences' raised by government authorities and environmental groups as early as the 1980s (Der Spiegel 1982). The state's environmental ministry suppressed a study warning of the disastrous ecological impacts of coal-mining in the region (including biodiversity loss, ecosystem degradation and desertification) and doubting RWE's ability to recultivate, and/or mitigate the impact of the mine (ibid.).

Green-washing activities take place on different levels of operation. Internationally, RWE has been leading efforts to 'improve' coal supply chains through *Better-coal*, a voluntary initiative that involves mine audits and stakeholder engagement, which serves as convenient opportunity to deflect critique, according to research participants (Brock 2018). Domestically, the company promotes its research on sustainable coal – framing its power plants as 'sustainable' due to achieved CO_2 emission reductions – and increases in efficiency as well as carbon offsetting. In its 'innovation centre', RWE publicly displays its testing of carbon capture and storages technologies ('CO_2 washing') as 'technology of the future', and 'almost ready for application'. To ensure further local support, RWE set up a recultivation centre that is responsible for its nature 'restoration' work, as part of (legally required) compensation measures in the form of enormous environmental-engineering experiments based on the belief in the human capacity to recreate nature (Brock 2020c). These offsets are meant to compensate for the destruction of species habitat (read non-human forest life) in the *Hambacher Forest*. Compensation measures include the newly restored *Sophienhöhe*,[12] an artificial, forested hill that was 'built from scratch' by the mine operator, according to RWE research participants. The offset involves careful planning and 'scientifically informed' mixing of soils – creating a diversity of ecosystems from more barren, sandy areas featuring 'rare species' to more fertile forest grounds that 'require' the continued destruction of the original forest to secure provision of topsoil (according to the company). Ironically, RWE not only built the largest artificial mountain in Europe but has simultaneously been financing mountaintop *removal* to mine coal in the United States (Hecking 2016). The new landscape or 'better nature', according to RWE, involves replanting trees and shrubs, establishing artificial bodies of water and 'local biodiversity hotspots' that are complete with resettled ant hills, relocated hazel dormouse colonies and dead tree trunks for breeding habitat. The restoration of *Sophienhöhe* is frequently showcased and has become a destination for regular scientific and touristic excursions and research projects. The creation of this 'better nature' is based on the very same violent processes of classification, quantification and measuring of life mentioned earlier – what Camila Moreno and others have called 'ecological epistemicide' (Moreno et al. 2015) – ignoring interconnections and social relations to the land and enabling claims of 'net gain' of trees (Brock 2020c).[13] In effect, *Sophienhöhe* is the outcome of RWE's efforts to make nature commensurable, legible and controllable; requiring continuous surveillance, monitoring and 'careful management' including regular fertilizer application for decades after planting these 'new forests'.

Sophienhöhe forms part of the spectacularization and commodification of the mining experience that is manifest in its transformation through 'communicative infrastructure' and into 'extractive attractions' (Brock & Dunlap 2018, 40; Brock 2020c). The *Sophienhöhe* contains 150 kilometres of hiking and cycling trails, an educational nature train for students, numerous visitor points including lookouts, Celtic tree circles and a 'giant redwood trail' to spectacularize the visit. Some trails are equipped with information boards containing QR codes to allow visitors to experience 'nature' through their smart phones and learn about 'the new landscape and its flora and fauna' (RWE Power 2016). Novel technology is thus used to mediate human relationships not only with the fauna and flora around them but also with their 'creator' – RWE. 'New nature' is heavily pre-structured and policed to prevent engagement beyond Sunday-strolling and dog walking, signposted and delineated by shrubs to keep people on the path: 'spatial environment[s] saturated with contemporary ideologies of containment and exclusion' (Ferrell 2012, 1688). Signs, rules, 'natural grids' and fellow visitors prevent exploration beyond the pre-planned trails, turning *Sophienhöhe* into a 'highly regulated, predictable and enclosed environment – like city parks positioned to serve as PR' (Brock & Dunlap 2018, 41).

Sophienhöhe is complemented by the creation of tourism opportunities around the mine, such as viewing platforms complete with commercial opportunities including a bar and a restaurant. Visitors are invited to enjoy the view over the mine from the revealingly named, *terra nova* ('New Earth' platform), modelled after a beach resort in anticipation of the planned transformation of the mine into Germany's second biggest lake upon mine closure.

> Visitors from near and far are invited to enjoy the view, drinks, food and games, and applaud the 200 plus-meter long diggers, the 'largest mobile machines of the world', invoking fantasies of huge playgrounds where soil is shifted and men have God-like control over both machinery and nature. (Ibid.)

Through the creation of 'better nature', its diversity of greening activities, and corporate social responsibility activities, RWE draws in conservation organizations and other potential critics to ensure the smooth functioning of the system (Brock 2018, 2020c). The goal of such corporate engagement is 'to isolate the radicals, cultivate the idealists and educate them into becoming realists, then co-opt the realists into agreeing with industry' (Lubbers 1999, n.p.). Conservationists are invited to the RWE recultivation conference, given a stage to present their research on orchids and butterflies, and receive public praise for their work. Local people are sent regular 'neighbourhood magazines', in which they can learn about RWE's recultivation work and the unruly and deranged 'radical' forest defenders. Other community engagement activities include RWE's 'baking cart' that drives across the country handing out baked goods, recipes, RWE material and energy-saving advice. The company was also engaged in a 'Peace Plan' as part of the 'Hambacher Dialogue' where it engaged with 'moderate' protesters, and has undertaken a large-scale acceptance study, *The Power of Participation*, to explore how stakeholder engagement and dialogue can 'avoid or reduce resistance' against megaprojects to protect 'the future viability of our business' (RWE AG 2012).

At the same time, coal-mining – and its social and ecological 'costs' – is further normalized through the capturing of hearts and minds of surrounding populations, planting pro-corporate ideologies, industrial desires, and fears of 'de-industrialization', 'blackouts' and 'primitivism' (Brock and Dunlap 2020). These are fostered by RWE's Public Relations work, lobbying and (so-called) Corporate Social Responsibility activities, complemented by its infiltration into decision-making bodies at all levels of the German government. 'Wherever decisions are taken, you find people who work for RWE or have worked for RWE', according to one local resident – testifying to RWE's role in shaping the physical, political, cultural and social environments in the Rhineland and beyond. These technologies serve to invisibilize the inherent violence in industrial coal-mining (or any large-scale electricity production) as well as the violence against forest defenders and dissidents (Brock and Dunlap 2020, 40 & 44).

Opposition against the mine has been harshly disciplined through various forms of (aggressive) policing, public–private security partnerships (Hissel 2015), surveillance, arrests and court procedures, subjecting land defenders and residents to ever greater control. More combative resistance against the mine has been met with police and corporate violence involving the increasing criminalization of forest defenders, physical attacks and threats of rape and death. The German state, of course, is intrinsically tied to fossil fuel interests, large-scale energy projects and infrastructure provision, having to defend such 'critical infrastructure' projects at all costs (Europol 2016, 8). It is no coincidence then that 'protests, vandalism, blockades and "lock-ons" ' against resource extraction companies and 'large-scale infrastructure' are singled out in Europol terrorism reports (Europol 2016, 43), branding anti-capitalist, animal, anarchist and environmental social movements as 'extremist' and 'terrorist' (ibid.). The mine is 'defended', however, not only by state/security forces and the media but also by all those who are captivated by ideas of progress, modernization and the green economy, having learned to hold dear the comforts gained and the 'promise' of good, 'honest' mining jobs.

Meanwhile RWE's sponsorships and multiple strategies to buy and engineer consent create new dependencies while the displacement of entire villages increases social fragmentation and alienation from each other and the land, breaking down social relationships. The world's largest hole continues to migrate. This 'hole' is visible from the four-lane highway that cuts through the landscape, allowing drivers to catch a glance of the moon-like landscape. The solar panels lining the highway, and

Figure 5.2 Windmills at the Edge of Hambach Mine. *Source:* Andrea Brock.

the enormous windmills around the mine, play into the 'greener future' that RWE promotes and markets in concert with the 'better nature' and 'pretty landscapes' the company claims to produce. The windmills become collaborators in the quest for accumulation and legitimacy, capturing the wind and feeding into the electric circuits which power the diggers 400 metres below. They illustrate the spectacular convergence of coal and renewables – the 'problem' and the 'solution' – for the sake of intensified industrial activity, economic growth and power (figure 5.2).

HARNESSING PEOPLE, CAPTURING WIND AND SUBDUING REBELLION IN OAXACA, MEXICO

The unique geographical features and positioning of the *Istmo* between the Gulf of Mexico and the Pacific Ocean have triggered a wind rush in the region (figure 5.3). It began with the 2003 USAID sponsored report, *Wind Energy Resource Atlas of Oaxaca* (Elliott et al. 2003), which mapped the 'excellent' wind sources in the region that the International Finance Corporation later called 'the best wind resources on earth' (IFC 2014, 1). The coastal *Istmo* can be divided into two sections: the North and the South. Sitting at the base of the Atravesada mountain range, the northern part of the region is generally regarded as Zapotec (*Binniza*), while the southern side is predominately Ikoot (Huave) territory. These territories overlap, while the Istmo is home to five different ethnic groups as well as a *mestizo* population (Campbell 1993)

Since 2004, wind energy development has resulted in the construction of 1,642 wind turbines (Rivas 2015; Rubí 2016) with twice this number being planned for

Figure 5.3   Wind Park in the North Istmo. *Source:* Wiki commons.

the region (Briseno 2016). While the desire for work, social development, and prosperity initially created support for wind projects in the region, many of these benefits remained unfulfilled or only benefited a minority of the population – politicians, their networks and select land owners (Dunlap 2017b). The towns and fishing communities of the 'South' witnessed wind park developments in the northern region, and as wind projects began spreading southward, people began organizing to resist them, especially those who valued their semi-subsistence lifestyle intertwined with the land and sea. Resistance and collaboration with the companies took on archetypal qualities in the Istmo (Borras et al. 2012; Hall et al. 2015). Contestation in the North is focused on exploitative land deals and labour contracts as locals fight for greater incorporation, as well as for individual and collective benefits. This includes unions – who were initially fighting for more wind parks – criticizing wind companies for bringing in technical employees and offering unequal pay between Mexican and Spanish workers. Meanwhile, in the South, the total rejection of wind energy projects largely arose, according to interviews, from the belief that wind companies (and the wider political system) cannot be 'trusted' since they 'propagate lies" to take people's land and "damage the sea" – thereby undermining local subsistence. Much has been published on wind energy in this region,[14] but here we highlight its role as an emerging apparatus of industrial control and vital usurpation.

The cries of the 1980s punk band *Oi Polloi*, 'Harness the wind – Harness the waves – We don't need this filthy nuclear waste!' has come to haunt the present. Emerging from the environmental movement as an alternative to coal and nuclear energy production (Stirling 2015), wind energy, especially its industrial-scale instillation, has been recuperated to renew business as usual. Until today, the environmental movement, leftist and other progressive circles view industrial-scale wind energy as a solution to the climate crisis and a pathway to ecologically sustainable futures. To lay bare the delusions of the green economy and the spell cast by renewable energy (marketing), the reality of wind energy development needs to be analysed for what it is actually doing in practice – rather than based on technological idealism or ecological modernization theory. This means briefly examining four aspects of wind energy development: the necessity of extractivism for raw materials, local social and ecological impacts, ownership and benefits, and wind power energy consumption.

The resources to create industrial-scale wind energy, first, come from mining and dredging of the Earth. Comprised of metals (iron, copper, aluminium, nickel, etc.), concrete, plastics, oil and rare earth minerals (dysprosium, praseodymium neodymium, terbium), so-called renewable wind energy requires not only traditional extractivism, road infrastructures, and (fossil fuel) transportation but also the deployment of marketing and security apparatuses to make extraction operations politically feasible (Downey et al. 2010; Bonds & Downey 2012). Highlighting this point early on, Eric Bond and Liam Downey recognize not only that increases in technological development can result in rising overall resource use but that "widespread commercialization of 'green' technologies has the potential to create new, more serious, or at least different environmental and humanitarian problems for less wealthy and less powerful groups" (Bond and Downey 2012, 181). While the ecological and policing cost of mining is well-documented (Brock & Dunlap 2018/2017; Guezuraga et al. 2012; Veltmeyer 2013; Geenen & Verweijen 2017), mineral extraction also leaves a daunting shadow over wind and other renewable energy technologies.

A two-megawatt wind turbine uses roughly 150 metric tonnes of steel for the rein-forced concrete foundations, 250 metric tonnes for the rotor hubs and nacelles and 500 metric tonnes for the tower (Smil 2016b). This also includes 3.6 tonnes of copper per megawatt (Smith 2014). Drawing on a World Bank report (La Porta et al. 2017), Jason Hickel estimates that to produce an annual output of about 7 terawatts of electricity by 2050 with wind and solar infrastructure will require mining '34 million metrics tons of copper, 40 million tons of lead, 50 million tons of zinc, 162 tons of aluminum, and no less than 4.8 billion tons of iron' (2019: n.p.). This estimate does not take into account fuels necessary for mining, processing, manufacturing and transporting raw materials and manufactured components. According to Begoña Guezuraga and colleagues the main contributors of wind energy's CO_2 footprint are steel, concrete and cast-iron production, while plastic production constitutes the most energy-intensive process (Guezuraga et al. 2012). The production of every ton of steel requires roughly 0.8 ton of coking (metallurgical) coal,[15] in addition to the energy required for steel *production*. While carbon accounting has surreptitiously justified these processes, the issue of mining and processing rare earth minerals to create wind turbine permanent magnet generators remains publicly neglected.

Baotou (Inner Mongolia) and South East China have historically produced between 85–98 per cent of rare earth metals used in wind turbines, electric cars, smart phones and other technologies (Hongiao 2016). Ninety-eight per cent of the heavy rare earth elements used in the EU came from China in 2020 (European Commission 2020). Between 2014 and 2017, according to Kalyeena Makortoff, 80 per cent of the US rare earth imports originated from China, who currently 'accounts for about 70% of global production' (Makortoff 2019). In a BBC report, the Baotou mining and processing area is described as 'hell on Earth', a terrifying dystopic industrial environment filled with pollution and cluttered with factories, pipelines, high-tension wires and artificial lakes filled with 'black, barely-liquid, toxic sludge' that 'tested at around three times background radiation' (Maughan 2015, 1). The reliance on Chinese resources and consequent fearmongering have recently led to EU and US strategies to diversify supply chains and push new extractive frontiers elsewhere. In response to China stopping a few shipments of rare earth minerals in 2012 in what was soon politically constructed to be a "supply crunch" triggering political panic and new investments across the world, attention was directed towards "strengthening the European rare earth supply chain" (Ahonen et al. 2014) and rare earths quickly "became 'strategic', and 'vital' materials crucial to 'security', 'technology', and 'the future'" (Klinger 2018).[16]

Mined through open-pit, underground and in-situ leaching methods (Haque et al. 2014), rare earth ore deposits contain 'low concentrations [of desired minerals] ranging from 10 to a few hundred parts per million by weight' and, especially in ion-adsorption clay, are 'symbiotic or associated with the radioactive elements uranium and thorium' (Yang et al. 2013, 133). Rare earth mining and processing, Nawshad

Haque and colleagues write, tend to be "energy, water and chemical intensive with significant environment risks affecting water discharges (radionuclides, mainly thorium and uranium; heavy metals; acid; fluorides), tailing management and air emissions" (Haque et al. 2014, 621). While rare earth elements are not actually rare at all, what is rare about them 'are the places where it is politically acceptable to mine and process them in a cost-effective manner' (Klinger 2015, 574).

Renewable energy thus involves socially and ecologically destructive mining processes with large amounts of tailings that contain heavy metals, toxic and radioactive wastes, which end up in the air, water, soil, animals and humans. Based on the same World Bank report, Hickel estimates there will be 35–70 per cent increase in neodymium – an essential mineral for wind turbines – and for grid battery storage over '40 million tons of lithium," which is a '2,700 percent increase over current levels of extraction (Hickel 2019).

The quantity and intensity of chemicals and toxic materials pouring into ecosystems are difficult to measure not only because of political but also epistemic reasons in accounting for full-spectrum environmental impacts. While in theory, Amory Lovins (2017) points out, wind turbines could be built without rare earth minerals with geared turbines, in practice this appears not to be the reality for industrial-scale wind parks – especially offshore wind parks and those in areas of extreme wind.[17] Like other industrial enchantments (such as computers and smart technologies), wind farms continue to require extractivism and generate toxic, radioactive and later, electronic waste. A '3.1 MW wind turbine created 772 to 1807 tons of landfill waste, 40 to 85 tons of waste sent for incineration and about 7.3 tons of e-waste per unit', explain Benjamin Sovacool and colleagues, who estimate that 1000,000 new wind turbines by 2050 to meet climate change mitigation standards 'will result in another 730,000 tons of e-waste' (Sovacool 2020, 1–19). Recycling capacities are low and varying between materials, yet retain roughly a 20 per cent recycling rate (ibid.), which the EU is currently trying to improve. Raw material extraction and e-waste are absent from much carbon accounting, and thus often invisible in the climate change debates.

Drawing on the experience of the Istmo, the second aspect of wind energy relates to the social and ecological impacts generated by wind turbines, the result of the placement, construction, and operation of wind parks. The *placement* of wind turbines requires locating suitable land and running tests akin to those published in the *Wind Resource Atlas of Oaxaca*. This necessitates negotiating not only the physical geography of hills, trees, bedrock and ground water but also the human geography of local political leaders, elites and landowners in the region. The land contracting is complicated by illegible and contested land relations, such as with *ejidos* and communal land.[18] Securing land in the Istmo requires various mechanisms, creating at times contradictory dynamics including limited or selective consultation and benefit sharing; neglecting economic, cultural and ecological impacts; rolling out wind company propaganda – or Public Relations – to parade ideas of jobs, individual prosperity and collective social development; and deploying manipulation, intimidation and deception tactics led by middlemen ('coyotes'), to secure land. Once land is secured, *construction* begins with the clearing of trees, bushes and other plants (including local herbs/medicines) to build roads, wind turbine foundations and subterranean and above the ground power lines. Digging wind turbine foundations requires holes that are roughly 7–14 metres (32–45 ft.) deep and about 16–21 metres (52–68 ft.) in diameter, depending on the specific geological composition of the land. These holes, as already mentioned, are filled with large amounts of steel and concrete. Notably, foundations are much deeper in areas without bedrock, such as the Lagoon Superior where local fishers claim that foundations were up to 70 metres deep. In the Istmo, fresh ground water is located 1 to 3 metres below the ground and wind turbine foundations which replace this water with steel-reinforced concrete foundations. Once in *operation*, killing of birds and other animals has been documented (Ledec et al. 2012), along with testimonies of oil leaking into the grazing grass and water wells. Alterations to the water table, the raising of roads and the constant swirling of the turbines, farmers report, cause extreme drying and flooding of the land. '[E]ven in this weather my tomato has gone dry – really fast. I am not going to be able to farm in the rainy season because of the road they made over there is seventy centimeters higher', explains a farmer, who compares their land with being 'inside a pool'.[19]

Other impacts have been reported in areas where wind turbines are built close to cities and bodies of water. In towns like La Ventosa, which is nearly enclosed by wind turbines and draped with electrical infrastructure, people report symptoms akin to the 'wind turbine syndrome' – headaches, tinnitus, insomnia, hypertension – and other severe illnesses (Pierpont 2009). While this is supported by a range of studies,[20] it requires further investigation. Wind energy development on and/or near the sea, as in the case of the Barra de Santa Teresa (Barra), digging, drilling and the use of heavy construction machinery have severely impacted aquatic populations that are extremely sensitive to electromagnetic currents and lights (Premalatha et al. 2014). Fishermen reported that aircraft warning lights (some that would even mimic a strobe light) from completed wind parks were pushing the fish farther away into the Lagoon Superior. For fishermen, this meant having to drive elsewhere to fish, which fermented, according to a local human rights activist, an 'inter-ethnic conflict' that was caused by the wind energy projects.[21] Residents from towns recognized locally as collaborating with and benefiting (however, contentious and disproportionate) from wind companies are now visiting other towns actively in resistance against the wind projects – and, consequently, without wind parks – to go fish, causing fights and conflicts to break out. This happened between as well as within towns (Dunlap 2018).

Wind park *ownership and local benefits* are heavily conditioned by neoliberal structural adjustment policies that favour national and foreign corporate acquisition. Wind parks are incentivized through green economy stimuli (grants and loans) coming from donor countries and private funds. Two funding sources are the Clean Development Fund (CDM), and the World Bank's Clean Technology and Climate Investment Funds (Dunlap 2014b, 2018/2017), which are linked to Certified Emission Reductions (CERs) for trading and speculation on the financial market. This connection to the market has been instrumental to the birth of the green economy.²² Wind companies are thus receiving increasing sums of money from public and private sectors to incentivize investment and profit making from wind park development, which is justified on the grounds of mitigating ecological and climate crises. This investment, however, is managed for profit maximization, turning climate change disaster into a new market opportunity, which becomes apparent when examining the use of wind energy in the chapter. Additionally, wind parks are operated by companies investing and working in other industries, such as Gas Natural Fenosa, a Spanish natural gas company that is the majority shareholder in the Bíi Hioxo wind park, which had been the source of immense conflict (Dunlap 2018/2017).²³

Finally, *what is wind energy used for?* Wind parks in the Istmo, based on the 1992 electricity law, are formally registered as 'self-supply' (*autoabastecimiento*) (Dunlap 2019). Self-supply electricity is generated privately and reserved for the investors or co-owners of wind parks, which are transporting electricity on public infrastructure from the Istmo to Guatemala, Belize, the United States and industrial areas within Mexico. Wind energy thus powers industrial construction companies (e.g., Cementos-Moctezuma, CEMEX), food processing corporations (Grupo Bimbo, Coca Cola), superstores (Walmart, Tiendas Chedrahui), and mining enterprises (Peñoles, Grupo Mexico) among others, rather than being used by the people living surrounded by or near these wind projects. Recently, after nine months of protest and deliberation between companies and local elites, it was agreed that Eólica del Sur would pay for three community wind turbines, finance a community centre, and pay 65 million pesos in taxes (Contreras 2018).

Wind energy is supporting and expanding conventional fossil fuel-based industrial activities, not transitioning away from them. Yet, environmental activists continue to cling onto renewable energy development in hope of creating an ecologically sustainable future. We argue, however, that industrial-scale, corporate-controlled wind energy production is captured by the capitalist grid that sustains and propels industrial growth and degradation, instead of replacing ecologically destructive modes of production and consumption. Investments in wind energy to 'offset' environmental damage continue to renew the images and degrading operations of industrial construction companies, food processing, superstores and mining companies, which feeds into the myths of 'sustainable mining', 'green uranium' and 'sustainable development' in general.²⁴

Currently, the 'sustainable' possibilities of wind energy have been eliminated by their operational scale, which reflects not only the existing energy-intensive infrastructure of industrial systems but also capitalist growth imperatives. Marketing and Public Relations campaigns – 'green-washing' – invisibilize this expansion and distract from the corporate growth and profit maximization imperatives that legally force companies to acquire increasing amounts of energy and natural resources. The latter contradicts and undermines the foundations of renewable energy transitions. Industrial-scale wind energy as we know it, along with its positive marketable vision, could not exist without the brutal and flagrant eradication of entire bioregions via the extraction of iron, copper, coal and other fossil fuel resources – often in countries of the Global South. Wind energy thus not only masks the flagrant destruction of mining metals, oil and rare earth minerals, reinforcing (neo)colonial trade links but these ecological damages also remain hidden behind uncritical notions of 'carbon accounting', 'just transition' and, in some cases, 'climate justice' as there is a lack of critical engagement with renewable energy infrastructure.

Wind energy turbines appear to be a less abrasive imposition, compared with coal mines or power plants, even at times when they surround entire towns (as in La Venta and La Ventosa) and are mixed with farming practices. This image conceals the global commodity chain and lifecycle on which they are dependent. While in theory, wind energy is 'renewable', and thus infinite, this framing hides two important facts. First is the limitation of their sustainability due to the need to replace wind turbines every thirty to forty years (Guezuraga et al. 2012), and second is the mineral and fossil fuel extractivism that is necessary for large-scale application of wind energy, which requires large amounts of steel, concrete, copper and rare metals. This is why renewable energy should more accurately be named fossil fuel+ (Dunlap n.d).

The present use of renewable energy, wedded to capitalist growth imperatives and powering 'dirty' industries, tears up, dominates and reconfigures the Earth in the image of industrial infrastructure, urbanization and, likely, 'nature reserves' – a dream long theorized by many enlightenment philosophers (Merchant 1983; Romanyshyn 1989; Adams 2014). The more people consent to the industrial regime and continue romanticizing renewable technologies, the more this dystopian project advances towards total environmental control.

REBRANDING EXTRACTION: THE RENEWABLE ENERGY–EXTRACTION NEXUS

The German Rhineland and the *Istmo* in Mexico are two very different places, but both are experiencing a type of natural resource extraction, sharing experiences and problems in oddly similar ways. A number of notable commonalities and differences between coal and wind energy extraction emerge. The *differences* are fairly obvious: geographic location, cultural context, processes of natural resources extraction – mining (coal) versus capturing (wind) – Intensities of extractive violence, and 'greening' activities/processes deployed. The *commonalities* are more interesting to unpack.

Both case studies present large-scale industrial developmental projects, or 'interventions', that directly rely on extractive industries and fossil fuels at different stages of their globalized supply chains including machinery, technologies and raw materials. People, villages and non-human habitats were regimented and destroyed to create coal mines and wind turbines. Both necessitate the same industrial infrastructure – transport, electricity and communication. As such, rather than challenge degrading industrial development and processes of capital accumulation, they secure hierarchical power relationships and corporate control over (human and non-human) nature. Both projects disproportionately benefit a political economic elite (shareholders and executives), with financial and ideological support from the public sector and large parts of the 'public'. State support is inherent to extractivism, itself a colonial ideology bound to state power (Acosta 2013). Indeed, state power itself has historically been – and continues to be – built on processes of extraction (energy production) and associated violence to control ecosystems and populations. Public support is secured through various social (counterinsurgency) technologies to co-opt, manage and pacify opposition to state-corporate agendas (Brock & Dunlap 2018). They involve investing into Public Relations, Corporate Social Responsibility programmes and public–private security partnerships to secure operations in the face of social fragmentation, environmental degradation and popular protest. Anarchist political ecology not only remains foundational in challenging the myths of (eco-/ neo) liberalism and nation state development but acknowledges the systemic problems of hierarchy, extreme divisions of labour and (malicious) competition – rotten relationships – and recognizes the viral and recuperative approaches to manufacturing social consent.

The 'management' of the various social and ecological impacts serves to hide the political and extractive violence inherent in both the Rhineland and the Istmo. The magnitude and implications of extractive violence are immediately clear with coal-mining – ancient forests, wetlands and grasslands full with human and non-human life are transformed into giant holes, moon landscapes and leaching ponds, causing displacement, degradation and death. Wind energy, at first sight, appears 'clean' in comparison, with shining metal towers and no noticeable emissions, standing above a landscape causing seemingly no disruptions (with the exception of dripping oil and bird corpses surrounding them). The negative ecological impacts are abrasive during the construction phase; strategic and (relatively) limited compared to mining. The problems with wind turbines are often related to scale, quantity and placement (e.g., distance from houses and sea life) of turbines, neglect of bird and animal mitigation strategies, as well as energy *usage* and decommissioning. The result is a type of 'slow extractive violence' that is steady and subtle (Nixon 2011). As discussed earlier, however, the real extractive violence with wind energy is concealed and exported out of sight and out of mind, not only in relatively isolated rural regions in the Global North but also in the Global South, which maintains fewer enforced environmental and human rights regulations (Szablowski 2007). By concealing extractive activities needed to construct wind turbines, colonial relations manifest in the export of politically violent and ecologically damaging extractive activities to the Global South. The latter enables greater acceptance and complicity among environmentalist, leftists and other 'progressives' who would (hopefully) otherwise condemn this resource colonialism and unequal ecological exchange.[25] Centre–periphery dynamics, with all of their nuances, still lurk in shadows of wind turbines and other renewable technologies.

The natural resource extraction sites in Germany and Mexico are linked through complex greening activities to legitimize operations and impacts. EU legislation requires biodiversity offsets to compensate for the ecological impacts of renewable energy projects on protected habitat. Meanwhile, in the Global South, renewable energy projects constitute offsets in and of themselves. In effect, this further ties nature protection to degradation, and links climate harm through industrial development in the Global North to renewable energy 'interventions' in the South. The latter thus serve to legitimize, and depoliticize, industrial operations in the North. German coal mine operator RWE not only engages in carbon offsetting in the Global North and South but also provides biodiversity offsets for German construction projects, selling 'eco-points' generated through nature 'restoration' to German municipalities (Hupp 2016; Brock 2020c). In the case of coal, we see a rebranding of the 'old' fossil fuel regime through such offsetting activities and promises of carbon capture and storage, fabricating the idea of 'sustainable coal' based on emission reductions and increases in efficiency.

Alternatively, in the wind energy case we see the marketing of a 'new' renewable energy regime. Yet, in their life cycles, at different (and multiple) points in their supply chains, both the extraction of coal and the production of wind energy retain a high level of socio-ecological disruption and/or destruction. This situation led Alexander Dunlap to argue that '[a]t best the dichotomy between fossil fuels and renewable energy is surreptitiously misleading and at worst it is a false dichotomy' (Dunlap 2017a, 257; see also Kisrch 2010, 2014; Sullican 2013; Brock & Dunlap 2018/2017). A comprehensive comparison of the destructive impacts from coal and wind, taking into account entire commodity chains of extractive machinery used in extractive sites, the mining operations themselves (coal, copper, rare earth, etc.), labour, processing of raw materials, transport, operation, decommissioning and overall life cycle is still lacking and needs investigation. Anarchist political ecology helps recognize the various and interrelated oppressions emerging from energy infrastructure, which includes acknowledging the social engineering and marketing of these projects. 'Greening' is being used in both sites to gain legitimacy, pacify dissent, and continue business as usual. This greening represents governmentality or 'eco-governmentality' (Ulloa 2013/2005), another weapon, or social technology of governance (Brock 2018), in the toolbox of governments, corporations and police–military practitioners – counterinsurgents – to manage rural protest and resistance led by indigenous people (Brock and Dunlap 2018; Dunlap 2017a) but also urban people protesting and ready to take action to create systemic change for ecologically just futures.

Both energy technology systems are further linked through the actors and interests behind them. The industrial processing facilities in Asia, Africa and Latin America associated with and powered by the fossil fuel economy are themselves producing essential components for cars, smart phones and industrial-scale wind turbines (Maughan 2015; Haque et al. 2014). Fossil fuel and mineral companies such as RWE, Gas Natural Fenosa, Grupo Mexico, or Peñoles buy or construct their own wind companies and industrial parks, using this energy to expand their operations or to create consortiums with industrial construction companies, (junk) food processing companies, superstores and mining companies, depriving local people of their resources. Not only does this resource colonialism retain a centre–periphery dynamic in the securing of raw materials but also in the *operation* and *use* of wind resources. After being exported out of their region, energy is converted and ultimately sold back to local people in form of processed goods (such as plastic), or in the form of infrastructural projects. Meanwhile, in Germany, RWE is investing in wind and solar installations which not only feed into the same grid that is powering coal extraction in the Rhineland but communicates 'sustainability' and social progress to the population (Brock 2020c). The Rhineland and the Istmo demonstrate the renewable energy–extraction nexus, which – instead of questioning the destructive trajectory of capitalist industrial progress – merges the 'normal' extractive and the 'green' economy to reinforce each other to continue feeding industrial expansion both materially and financially (Hildyard 2016). This nexus represents an intimate connection between conventional and renewable energy that not only share the same industrial lineage and technological continuum but are connected across different sites through the use and extraction of raw materials, companies and grid networks.

Anarchists know that the state and corporate entities are based on hierarchical ordering and social and ecological degradation, and their organizational existence and/or imperative are inseparable from their destructive behaviour – 'green' or otherwise. Anarchist political ecologists know that de-growth is a necessity and that relationships built on hierarchy, divisions of labour, commodification and exchange are doomed to redress the system-wide issues and traumas but instead advance agendas of control through the militarization and marketization of everything, everywhere and by every means. In the end, whether fossil fuel or wind energy, the industrial machine expands its infrastructural and fibre optic tentacles, violence and enchantment, as policymakers and the public alike tell themselves that climate change, biodiversity loss and ecological degradation are being mitigated (and now adapted to), while the production of industrial waste and economic growth continue.

CONCLUSION

Fossil fuels and renewables continue to be framed as 'good' and 'bad', 'clean' and 'dirty' or as 'the problem' and 'the solution', even by 'progressives' and environmental justice activists. Outside popular and media discourses, numerous scholars offer greater specification in types of 'renewable' energy generation regimes and the numerous political challenges confronting their participatory and equitable development (Burke & Stephens 2018; Newell 2019; Naumann & Rudolph 2020; Brock, Sovacool & Hook forthcoming). Within the literature on energy transitions and renewables, however, there remains a strong blind spot, ignoring the murky reality and rippling effects behind the raw material supply and processing webs of so-called 'renewable' and 'fossil fuel+' energy generation technologies[26]. Said differently, the unsavoury reality of the renewable energy–extraction nexus remains neglected. In this chapter, employing an anarchist political ecology lens to examine the rippling effects of socio-ecological oppression, we argue that this division is surreptitious and dangerous because it makes the degradation inherent in the contemporary 'green' industrial system invisible by hiding how it is actually *renewing destruction*. Instead, we argue, fossil fuels and industrial-scale, corporate-controlled renewables constitute two sides of the same coin – inseparable in terms of finance and profits, actors involved, power relationships surrounding and linking these technologies, corporate visions, energy uses and resulting inequalities. It is these same actors that are involved in both processes, linked not only through complex investment and finance networks and ownership patterns but also through physical processes, dependencies and shared supply chains. The alleged 'sustainability' of fossil fuel+ system necessitates the transformation of environments, fauna, flora and human life, causing physical, cultural and social disruption, degradation and destruction. Sustainability, for corporations, governments and many NGOs, not only refers to financial sustainability but also to the management of popular dissent and insurrection against the commodification, transformation and/or destruction of human and non-human lives. While 'the management of environmentalists is central to environmental management' (Levy 1997, 126–147), the green economy, as a component of the renewable energy–extraction nexus, is also about extracting, and deriving value from what environmentalists value the most: the process of so-called 'greening' itself. In sum, the sustainability employed by the green economy is about sustaining the arrogant and imbecilic direction of capitalist development, which includes developing green commodities and markets to the detriment of habitats, ecosystems and the climate.

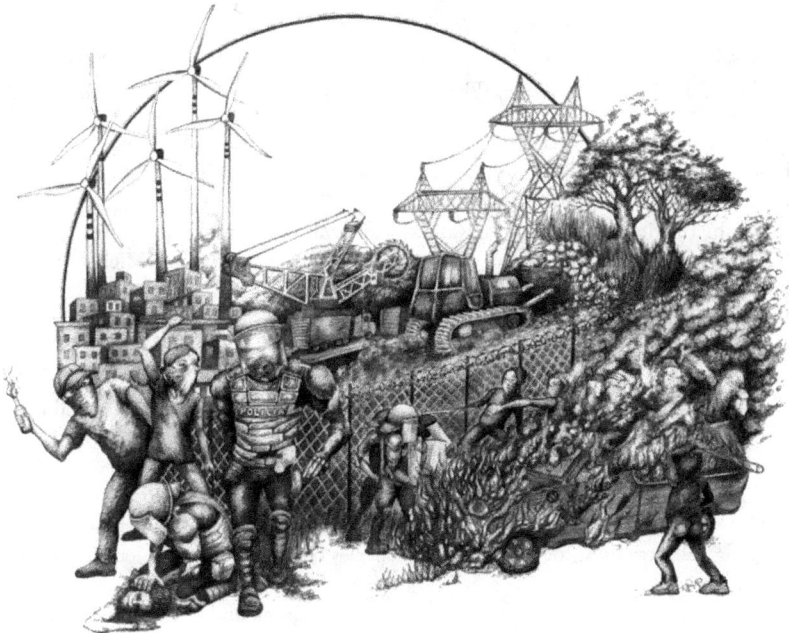

Figure 5.4   Land Defenders Continue. *Source:* Art by Riona'O Regan.

In this chapter, we hope to have redirected critical attention to the roots of the multiple social and ecological crises that are intertwined not (just) in tonnes of carbon (emissions) and industrial waste but in issues of power and control. The power and control exerted over ecosystems, animals and people to enforce a mode of industrial and computational development are causing various forms of environmental degradation, social discontent and disease. Let this chapter demonstrate that the green economy and its emerging instruments for both 'old' and 'new' energy systems demonstrate a continuation of 'war by ecological crisis' to control both human and non-human resources. We hope environmentalists, academics and others will begin to acknowledge this – with all the difficulties and depths that entails – when confronting and examining the process of techno-industrial development (figure 5.4).

NOTES

1. For examples see Greenpeace n.d.; European Commission 2019; 350.org 2020. For exceptions, see London Mining Network 2019; Sovacool et al., 2020.

2. Fieldwork in Germany included twenty-two semi-structured and countless informal interviews conducted between October 2016 and April 2017, building on long-term involvement with resistance movements against RWE's mining activities. Research in Mexico is based on 123 recorded semi-structured interviews conducted between December 2014 and May 2015, which also included a commitment to the collective resistance movements in the *Istmo* region.

3. Three archetypes of justifying atrocity: (1) subordination to 'the higher good' of the nation, the company; (2) discourses of 'savages', 'racially inferiority', 'poverty' etc.; and (3) positioning 'us', the church, state, company, etc., as 'saving', 'helping', 'civilizing' or 'educating' them – the 'Other'.

4. For details of harnessing Indigenous labour in Colombia, see Taussing (1987).

5. See Veblen 2009/1899; Horkheimer & Adorno 2002/1944; Bernays, 1947; Dugger 1989; Herman, & Chomsky 2010/1989.

6. For the continued relevance of Nazi laws around coal-mining, see Michel (2005).

7. For the continued relevance of Nazi laws around coal-mining, see Michel (2005).

8. See https://www.youtube.com/watch?v=SyrCiq_pQuo

9. The company is further involved in the privatization and operation of municipal electricity, gas and water distribution networks, street lighting systems and other local service provision. See RWE Group (2015, 89).

10. Minister president Armin Laschet in Tagesschau.de, Kraftwerk Datteln soll bis 2038 laufen, 30 December 2020. https://www.tagesschau.de/wirtschaft/unternehmen/uniper-datt eln-kohleausstieg-co2-laschet-101.html

11. For more examples and further analysis see Brock and Dunlap (2018/2017).

12. Other offset measures are the newly created 'bat-highways' that are meant to serve as navigating infrastructure for threatened bat species to facilitate their relocation into other pieces of forests.

13. RWE interviews and PR material, for details see Brock and Dunlap (2018/2017).

14. For examples, see Oceransky 2011; Juárez-Hernández & León 2014; Howe & Boyer 2015; Friede et al. 2017.

15. See https://www.letstalkaboutcoal.co.nz/future-of-coal/making-steel-without-coal/ and Diez et al. (2002).

16. Rare earth minerals and their geological knowledge production, Julie Klinger argues, have always been politically entangled and deeply colonial, imperial and militaristic. Rare earth elements became key to industrial and military development from the end of the nineteenth century onwards, and British, Austrian and German companies quickly came to dominate production, primarily in India and Brazil. See Klinger 2015 for their political, imperial, colonial and militaristic entanglements.

17. For a detailed and refreshing discussion in the Dutch context see Kiezebrink et al. (2018).

18. The *ejido* emerges from Article 27 of the 1917 Constitution, which provided land for farmers to use but not to buy and sell. After the 1992 alterations to Article 27 and the December 2013 Energy and Utility Act, land was allocated for residential and agricultural use and was governed by local assemblies made up of recognized community members. Article 27 still gave the Mexican state the right to resources underneath the topsoil and to control the land. *Ejidos* in Istmo are different from communal land, land governed by the community. Communal land (social property) is held collectively or shared communally, has no formal land title and does not have the same level of state involvement and control as *ejidos*.

19. Interview, 13 March 2015.

20. See Havas et al. 2011; Chapman 2012; Jeffery et al. 2014; Premalatha et al. 2014.

21. Interview, 21 March 2015.

22. See Dunlap & Fairhead 2014; Hunsberger et al. 2017; Fairhead et al. 2013; Corson et al. 2013; Dunlap 2019; Lohman 2008; Sullican 2010, 2013a, 2017.

23. Interview, 21 March 2015.

24. See Kirsch 2010 2014; Sullivan 2013b; Brock and Dunlap 2018/2017.

25. On unequal ecological exchange, see Hornborg 1998.

26. While the multi-scalar and industry connections of conventional and green extractivist projects require further scrutiny, important recent examples are Selwyn 2020; Hund et al. 2019.

REFERENCES

350.org. (2020). *Renewable Energy in Africa: An Opportunity in a Time of Crisis.* https://7lo0w1yurlr3bozjw1hac3st-wpengine.netdna-ssl.com/files/2020/07/Renewable-energy-in-Africa-report-June-2020-screen.pdf.

Acosta, A. (2013). Extractivism and Neoextractivism: Two Sides of the Same Curse. *Beyond Development: Alternative Visions from Latin America*, 1: 61–86.

Adams, R. E. (2014). Natura Urbans, Natura Urbanata: Ecological Urbanism, Circulation, and the Immunization of Nature. *Environment and Planning D: Society and Space* 32, no. 1: 12–29.

Ahonen, S., Arvanitidis, N., Auer, A., Baillet, E., Bellato, N., Binnemans, K., Blengini, A., Bonato, D., Brouwer, E., Brower, S., Buchert, M., & Reinhard. (2015). *Strengthening the European Rare Earths Supply-Chain: Challenges and Policy Options.* European Rare Earths Competency Network (ERECON). Working Papers cea-01550114, HAL.

Alexander, B. K. (2008). *The Globalization of Addiction: A Study in Poverty of the Spirit.* New York: Oxford University Press.

Anonymous. (2016). *Text Concerning Hambach Forest (Germany)*. https://325.nostate.net/wp-content/uploads/2016/06/return-fire-vol3-contents.pdf.

Avila-Calero, S. (2017). Contesting Energy Transitions: Wind Power and Conflicts in the Isthmus of Tehuantepec. *Journal of Political Ecology* 24: 993.

Bakunin, M. (2005/1871). God and the State. In *No Masters, No God*, edited by D. Guérin. pp. 150–152. Oakland: AK Press.

Beckert, S. (2014a). *Empire of Cotton: A Global History*. New York: Alfred A. Knopf.

———. (2014b). Slavery and Capitalism. *The Chronicle Review*. https://www.chronicle.com/article/slavery-and-capitalism/.

Berman, E. (1983). *The Ideology of Philanthropy: The Influence of the Carnegie, Ford, and Rockefeller Foundations on American Foreign Policy*. Albany: State University of New York Press.

Bernays, E. (1947). The Engineering of Consent. *The Annals of the American Academy of Political and Social Science* 250, no. 1: 113–120.

Bhambra, G. K. (2020). Colonial Global Economy: Towards a Theoretical Reorientation of Political Economy. *Review of International Political Economy* 28, no. 2: 307–322.

Böhm, S., & Dabhi, S. (Eds.). (2009). *Upsetting the Offset: The Political Economy of Carbon Markets*. London: MayFly.

Bonds, E., & Downey, L. (2012). Green Technology and Ecologically Unequal Exchange: The Environmental and Social Consequences of Ecological Modernization in the World-System. *Journal of World-Systems Research* 18, no. 2: 167–186.

Bonneuil, C., & Fressoz, J. B. (2016). *The Shock of the Anthropocene: The Earth, History and Us*. Verso Books.

Borras, S. M., Kay, C., Gómez, S., & Wilkinson, J. (2012). Land Grabbing and Global Capitalist Accumulation: Key Features in Latin America. *Canadian Journal of Development Studies /Revue canadienne d'e´tudes du developpement* 33, no. 4: 402–416.

Briseno, P. (2016). Frustran Extorsión a Empresa Española; Le Pedían $500 Mil. *Excelsior*. https://m.e-consulta.com/medios-externos/2016-09-24/frustran-extorsion-empresa-espanola-le-pedian-500-mil/.

Brock, A. (2018). *Conserving Power: An Exploration of Biodiversity Offsetting in Europe and Beyond*. Doctoral Thesis. University of Sussex.

———. (2020a). Enforcing Ecological Catastrophe at All Costs. *New Internationalist*. https://newint.org/features/2020/11/19/enforcing-ecological-catastrophe-all-costs.

———. (2020b). 'Frack off': Towards an Anarchist Political Ecology Critique of Corporate and State Responses to Anti-Fracking Resistance in the UK. *Political Geography* 82: 102246.

———. (2020c). Securing Accumulation by Restoration – Exploring Spectacular Corporate Conservation, Coal Mining and Biodiversity Compensation in the German Rhineland. *Environment and Planning E: Nature and Space* 251484862092457: 1–32.

Brock, A., & Dunlap, A. (2015). 'Love for Sale': Biodiversity Banking and the Struggle to Commodify Nature in Sabah, Malaysia. *Geoforum* 65: 278–290.

———. (2018/2017). Counterinsurgency for Wind Energy: The Bíi Hioxo Wind Park in Juchitán, Mexico. *The Journal of Peasant Studies* 45, no. 3: 630–652.

———. (2018). Normalising Corporate Counterinsurgency: Engineering Consent, Managing Resistance and Greening Destruction Around the Hambach Coal Mine and Beyond. *Political Geography* 62: 33–47.

Brock, A., Sovacool, B. K., & Hook, A. (2021). Volatile Photovoltaics: Green Industrialization, Sacrifice Zones, and the Political Ecology of Solar Energy in Germany. *Annals of the American Association of Geographers*, 111(6), 1756–1778.

Burke, M. J., & Stephens, J. C. (2018). Political Power and Renewable Energy Futures: A Critical Review. *Energy Research & Social Science* 35: 78–93.

Büscher, B., & Davidov, V. (Eds.). (2013). *The Ecotourism-extraction Nexus: Political Economies and Rural Realities of (Un)comfortable Bedfellows*. London: Routledge.

Campbell, H., Binford, L., Bartolomé, M., & Barabas, A. (1993). *Zapotec Struggle: Histories, Politics, and Representations from Juchitán, Oacaca*. Washington: Smithsonian Institution Press.

Chapman, S. (2012). The Sickening Truth about Wind Farm Syndrome. *New Scientist* 216, no. 2885: 26–27.

Clark, J. (2020). What is Eco-Anarchism. *The Ecological Citizen* 3: 9–14.

Clegg, J. (2015). Capitalism and Slavery. *Critical Historical Studies* 2, no. 2: 281–304.

Contreras, G. A. T. (2018). Wind Energy Development in Mexico: An Authoritarian Populist Development Project? *Transnational Institute*. https://www.tni.org/files/article-downloads/erpi_cp_19_contreras.pdf.

Corson, C., MacDonald, K. I., & Neimark, B. (Eds.). (2013). Grabbing 'Green': Markets, Environmental Governance and the Materialization of Natural Capital, Special Issue. *Human Geography* 6, no. 1: 1–15.

Crimethinc. (n.d.). From Democracy to Freedom. *Crimethinc*. http://crimethinc.com/texts/r/democracy/.

Cullather, N. (2006). The Target Is the People: Representations of the Village in Modernization and U.S. National Security Doctrine. *Culture Politics* 2, no. 1: 29–48.

———. (2013/2010). *The Hungry World: America's Cold War Battle against Poverty in Asia*. Cambridge: Harvard University Press.

Dalakoglou, D., & Harvey, P. (2012). Mobilities Special Issue: Roads and Anthropology. *Mobilities* 7, no. 4: 459–465 .

Davis, R., & Zannis, M. (1973). *The Genocide Machine in Canada*. Montreal: Black Rose Books.

Debord, G. (1967). *The Society of the Spectacle*. New York: Zone Books.

Der Spiegel. (1982). *Das größte Loch*. http://www.spiegel.de/spiegel/print/d-14356858.html.

Diez, M. A., Alvarez, R., & Barriocanal, C. (2002). Coal for Metallurgical Coke Production: Predictions of Coke Quality and Future Requirements for Cokemaking. *International Journal of Coal Geology* 50, no. 1–4: 389–412.

Downey, L., Bonds, E., & Clark, K. (2010). Natural Resource Extraction, Armed Violence, and Environmetnal Degradation. *Organization Environment* 23, no. 4: 453–474.

Dugger, W. (1989). *Corporate Hegemony*. Connecticut: Greenwood Press.

Dunlap, A. (n.d.). End the 'Green' Delusions: Industrial-Scale Renewable Energy Is Fossil Fuel+. *Verso Blog*. https://www.versobooks.com/blogs/3797-end-the-green-delusions-industrial-scale-renewable-energy-is-fossil-fuel.

———. (2014a). Power: Foucault, Dugger and Social Warfare. In *The Bastard Chronicles: Social War*, edited by BASTARD Collective. Berkeley: Ardent Press: 55–106.

———. (2014b). Permanent War: Grids, Boomerangs, and Counterinsurgency. *Anarchist Studies* 22, no. 2: 55–79.

———. (2017a). From Primitive Accumulation to Modernized Poverty: Examining Flush-Toilets through the Four Invaluation Processes. *Forum for Social Economy* 47, no. 2: 1–21.

————. (2017b). *Renewing Destruction: Wind Energy Development in Oaxaca, Mexico.* Doctoral Thesis. Amsterdam: Vrije Universiteit University.

————. (2017c). 'The Town Is Surrounded:' From Climate Concerns to Life under Wind Turbines in La Ventosa, Mexico. *Human Geography* 10, no. 2: 16–36.

————. (2018a). The 'Solution'is Now the 'Problem:' Wind Energy, Colonisation and the 'Genocide-Ecocide Nexus' in the Isthmus of Tehuantepec, Oaxaca. *The International Journal of Human Rights* 22, no. 4: 550–573.

————. (2018b). Insurrection for Land, Sea and Dignity: Resistance and Autonomy against Wind Energy in Álvaro Obregón, Mexico. *Journal of Political Ecology* 25: 120–143.

————. (2019). *Renewing Destruction: Wind Energy Development in Oaxaca, Mexico.* London: Rowman & Littlefield.

————. (2020). The Politics of Ecocide, Genocide and Megaprojects: Interrogating Natural Resource Extraction, Identity and the Normalization of Erasure. *Journal of Genocide Research* 23, no. 2: 212–235.

Dunlap, A., & Fairhead, J. (2014). The Militarisation and Marketisation of Nature: An Alternative Lens to 'Climate-Conflict'. *Geopolitics* 19, no. 4: 937–961.

Dunlap, A., & Jakobsen, J. (2020). *The Violent Technologies of Extraction: Political Ecology, Critical Agrarian Studies and the Capitalist Worldeater.* London: Palgrave.

Dunlap, A., & Sullivan, S. (2019). A Faultline in Neoliberal Environmental Governance Scholarship? Or, Why Accumulation-by-Alienation Matters. *Environment and Planning E: Nature and Space* 3, no. 2: 552–579.

Edward, E. B. (2014). *The Half Has Never Been Told: Slavery and the Making of Modern Capitalism.* New York: Basic Books.

Elliott, D., Schwartz, M., Scott, G., Haymes, S., Heimiller, D., & George, R. (2003). *Wind Energy Resource Atlas of Oaxaca* (No. NREL/TP-500-34519). Golden, CO: National Renewable Energy Lab (NREL).

Ellul, J. (1964/1954). *The Technological Society.* New York: Vintage Books.

Europol. (2016). European Union Terrorism Situation and Trend Report 2016. *Europol.* http://statewatch.org/news/2016/jul/europol-te-sat-repor-2016.pdf.

European Commission. (2019). *Clean Energy for all Europeans.* Directorate-General for Energy. https://op.europa.eu/en/publication-detail/-/publication/b4e46873-7528-11e9-9f05-01aa75ed71a1/language-en?WT.mc_id=Searchresult&WT.ria_c=null&WT.ria_f=3608&WT.ria_ev=search.

————. (2020). Communication from the Commission to the European Parliament, the Council, the European Economic and Social Committee and the Committee of the Regions. *Critical Raw Materials Resilience: Charting a Path towards Greater Security and Sustainability.* COM/2020/474 final, 3-9-2020 EU. https://eur-lex.europa.eu/legal-content/EN/TXT/?uri=CELEX:52020DC0474.

Fairhead, J., Leach, M., & Scoones, I. (2012). Green Grabbing: A New Appropriation of Nature? *Journal of Peasant Studies* 39, no. 2: 237–261.

Ferrell, J. (2012). Anarchy, Geography and Drift. *Antipode* 44, no. 5: 1687–1704 .

Fitzpatrick, B. (2018). *Corrosive Consciousness.* Jacksonville: Enemy Combatant.

Friede, S. (n.d.). *Enticed by the Wind: A Case Study in the Social and Historical Context of Wind Energy Development in Southern Mexico.* Washington, DC: The Wilson Center. https://www.wilsoncenter.org/publication/enticed-the-wind-case-study-the-social-and-historical-context-wind-energy-development.

Geenen, S., & Verweijen, J. (2017). Explaining Fragmented and Fluid Mobilization in Gold Mining Concessions in Eastern Democratic Republic of the Congo. *The Extractive Industries and Society* 4, no. 4: 758–765.

Gelderloos, P. (2013). *The Failure of Nonviolence: From Arab Spring to Occupy.* Seattle: Left Bank Books.

———. (2017). *Worshiping Power: An Anarchist View of Early State Formation.* Oakland: AK Press.

Gorz, A. (1980). *Ecology as Politics.* Boston: South End Press.

Graeber, D. (2006). Turning Modes of Production Inside Out: Or, Why Capitalism is a Transformation of Slavery. *Critique of Anthropology* 26, no. 1: 61–85.

Greenpeace. (n.d). *100% Renewable Energy For All.* https://www.greenpeace.org/usa/issues/re newable-energy/.Guezuraga, B., Zauner, R., & Pölz, W. (2012). Life Cycle Assessment of Two Different 2 MW Class Wind Turbines. *Renewable Energy* 37, no. 1: 37–44.

Güven, F. (2015). *Decolonizing Democracy: Intersections of Philosophy and Postcolonial Theory.* London: Lexington Books.

Hall, R., Edelman, M., Borras Jr, S. M., Scoones, I., White, B., & Wolford, W. (2015). Resistance, Acquiescence or Incorporation? An Introduction to Land Grabbing and Political Reactions 'From Below'. *Journal of Peasant Studies* 42, no. 3–4: 467–488.

Haque, N., Hughes, A., Lim, S., & Vernon, C. (2014). Rare Earth Elements: Overview of Mining, Mineralogy, Uses, Sustainability and Environmental Impact. *Resources* 3, no. 4: 614–635.

Havas, M., & Colling, D. (2011). Wind Turbines Make Waves: Why Some Residents near Wind Turbines Become Ill. *Bulletin of Science, Technology and Society* 31, no. 5: 414–426.

Hawkins, A. D., Roberts, L., & Cheesman, S. (2014). Responses of Free-Living Coastal Pelagic Fish to Impulsive Sounds. *The Journal of the Acoustical Society of America* 135, no. 5: 3101–3116.

Hecking, C. (2016). Das ist der Gipfel. *Die Zeit.* http://www.zeit.de/2016/13/rwe-usa-kohle -minen-sprengung-bergbau.

Herman, E. S., & Chomsky, N. (2010/1989). *Manufacturing Consent: The Political Economy of the Mass Media.* New York: Random House.

Hickel, J. (2019, Sept 6). The Limits of Clean Energy. *Foreign Policy.* https://foreignpolicy .com/2019/09/06/the-path-to-clean-energy-will-be-very-dirty-climate-change-renewables/.

Hildyard, N. (2016). *Licensed Larceny: Infrastructure, Financial Extraction and the Global South.* Manchester: Manchester University Press.

Hissel, Y. (2015). RWE liebt die Polizei. *taz.* http://www.taz.de/!5224546/.

Hongiao, L. (2016). The Bottleneck of a Low-Carbon Future. *Chinadialogue.* https://chinadi alogue.net/article/9209-The-bottleneck-of-a-low-carbon-future.

Horkheimer, M., & Adorno, T. W. (2002/1944). *Dialectic of Enlightenment: Philosophical Fragments.* Stanford Unviersity Press.

Hornborg, A. (1998). Towards an Ecological Theory of Unequal Exchange: Articulating World System Theory and Ecological Economics. *Ecological Economics* 25, no. 1: 127–136.

Howe, C., & Boyer, D. (2015). Aeolian Politics. *Distinktion: Scandinavian Journal of Social Theory* 16, no. 1: 31–48.

Huff, A., & Brock, A. (2017). Intervention – 'Accumulation by Restoration: Degradation Neutrality and the Faustian Bargain of Conservation Finance'. *Antipode Foundation.* https ://antipodefoundation.org/2017/11/06/accumulation-by-restoration/.

Hund, K., La Porta, D., Fabregas, T. P., Laing, T., & Drexhage, J. (2019). Minerals for Climate Action: The Mineral Intensity of the Clean Energy Transition. *The World Bank Group.* http://pubdocs.worldbank.org/en/961711588875536384/Minerals-for-Climate-Action-The-Mineral-Intensity-of-the-Clean-Energy-Transition.pdf.

Hunsberger, C., Corbera, E., Borras Jr, S. M., Franco, J. C., Woods, K., Work, C., … & Park, C. (2017). Climate Change Mitigation, Land Grabbing and Conflict: Towards a Landscape-Based and Collaborative Action Research Agenda. *Canadian Journal of Development Studies/Revue canadienne d'études du développement* 38, no. 3: 305–324.

Hupp, H. (2016). *Ökopunkte sind der falsche Weg.* http://piratenpartei-bruehl.de/2016/09/23/oekopunkte-sind-der-falsche-weg/.

IFC. (2014). Investments for a Windy Harvest: Ifc Support of Teh Mexican Wind Sector Drives Results. *International Finance Corporation.* Washington, DC: World Bank Group.

Ince, A., & De la Torre, G. B. (2016). For Post-Statist Geographies. *Political Geography* 55: 10–19.

James, C. L. R. (1938). *The Black Jacobins: Toussaint L'Ouverture and the San Domingo Revolution.* London: Secker and Warburg.

Jeffery, R. D., & Brett Horner, C. M. A. (2014). Industrial Wind Turbines and Adverse Health Effects. *Canadian Journal of Rural Medicine* 19, no. 1: 21.

Johnson, W. (2013). *River of Dark Dreams: Slavery and Empire in the Cotton Kingdom.* Harvard University Press.

Juárez-Hernández, S., & León, G. (2014). Wind Energy in the Isthmus of Tehuantepec: Development, Actors and Social Opposition. *Problemas del Desarrollo: Revista Latinoamericana de Economía* 45, no. 178: 1–9.

Karuka, M. (2019). *Empire's Tracks: Indigenous Nations, Chinese Workers, and the Transcontinental Railroad.* University of California Press.

Katsiaficas, G. (2006/1997). *The Subversion of Politics: European Autonomous Social Movements and the Decolonization of Everyday Life.* Oakland: AK Press.

Kiezebrink, V., Wilde-Ramsing, J., & Ten Kate, G. (2018). *Human Rights in Wind Turbine Supply Chains: Towards a Truly Sustainable Energy Transition.* SOMO and Actionaid. https://www.somo.nl/wp-content/uploads/2018/01/Final-ActionAid_Report-Human-Rights-in-Wind-Turbine-Supply-Chains.pdf.

Kirsch, S. (2010). Sustainable Mining. *Dialectical Anthropology* 34, no. 1: 87–93.

———. (2014). *Mining Capitalism: The Relationship between Corporations and Their Critics.* Berkeley: University of California Press.

Klinger, J. M. (2015). A Historical Geography of Rare Earth Elements: From Discovery to the Atomic Age. *The Extractive Industries and Society* 2: 572–580.

———. (2018). Rare Earth Elements: Development, Sustainability and Policy Issues. *The Extractive Industries and Society* 5: 1–7, 3.

La Porta, D., Hund, K., McCormick, M., Ningthoujam, J., & Drexhage, J. (2017). *The Growing Role of Minerals and Metals for a Low Carbon Future.* Washington, DC: The World Bank. http://documents.worldbank.org/curated/en/207371500386458722/The-Growing-Role-of-Minerals-and-Metals-for-a-Low-Carbon-Future.

Landstreicher, W. (n.d.). This is What Democracy Looks Like. *Elephant Editions.* https://theanarchistlibrary.org/library/various-authors-this-is-what-democracy-looks-like.

Lasswell, H. D. (1934). Propaganda. In *Encyclopedia of the Social Sciences.* edited by E. Seligman. New York: Macmillan: 524.

Ledec, G. C., Rapp, K. W., & Aiello, R. G. (2011). *Greening the Wind: Environmental and Social Considerations for Wind Power Development*. Washington, DC: The World Bank.

Levy, D. L. (1997). Environmental Management as Political Sustainability. *Organization & Environment* 10, no. 2: 126–147.

Li, T. M. (2011). Centering Labor in the Land Grab Debate. *Journal of Peasant Studies* 38, no. 2.

Lohmann, L. (2008). Carbon Trading, Climate Justice and the Production of Ignorance: Ten Examples. *Development* 51, no. 1: 359–365.

London Mining Network. 2019. *A Just(ice) Transition is a Post-Extractivist Transition: Centering the Extractive Frontier in Climate Justice*. https://londonminingnetwork.org/wp-content/uploads/2019/09/Post-Extractivist-Transition-report-2MB.pdf.

Lovins, A. (2017). Clean Energy and Rare Earths: Why Not to Worry. *Bulletin of the Atomic Scientists*. https://thebulletin.org/clean-energy-and-rare-earths-why-not-worry10785.

Lubbers, E. (1999). Field Report: Introducing the Panel Discussion of the Counter-Strategies Corporations Employ against Campaigns. *Cypersociology Magazine* 5. http://www.cybersociology.com/files/5_cyberstrategies.html.

Makortoff, K. (2019). US-China Trade: What are Rare-Earth Metals and What's the Dispute?. *The Guardian*. https://www.theguardian.com/business/2019/may/29/us-china-trade-what-are-rare-earth-metals-and-whats-the-dispute.

Maughan, T. (2015). The Dytopian Lake Filled by the World's Tech Lust. *BBC*. http://www.bbc.com/future/story/20150402-the-worst-place-on-earth.

McQueen, D. (2015). CSR and New Battle Lines in Online PR War: A Case Study of the Energy Sector and Its Discontents. In *Corporate Social Responsibility in the Digital Age*. edited by A. Adi, G. Grigore & D. Crowther. Bingley UK: Emerald Group Publishing Limited: 99–125.

Merchant, C. (1983). *The Death of Nature: Women, Ecology, and the Scientific Revolution*. New York: Harper & Row.

Michel, J. H. (2005). *Status and Impacts of the German Lignite Industry*. Göteburg: Secretariat on Acid Rain.

Moreno, C., Chassé, D. S., & Fuhr, L. (2015). *Carbon Metrics: Global Abstractions and Ecological Epistemicide*. Berlin: Heinrich Böll Stiftung.

Moses, A. D. (2002). Conceptual Blockages and Definitional Dilemmas in the 'Racial Century': Genocides of Indigenous Peoples and the Holocaust. *Patterns of Prejudice* 36, no. 4.

———. (Ed.). (2008). *Empire, Colony, Genocide: Conquest, Occupation, and Subaltern Resistance in World History, War and Genocide*. Oxford: Berghahn.

Mumford, L. (1967/1970). *Technics and Human Development: The Myth of the Machine*, vol. I (pp. 381–410). Harcourt Brace Jovanovich.

Naumann, M., & Rudolph, D. (2020). Conceptualizing Rural Energy Transitions: Energizing Rural Studies, Ruralizing Energy Research. *Journal of Rural Studies* 73: 97–104.

NDR. (2017). *G20-Krawalle*. http://www.ndr.de/nachrichten/hamburg/G20-Krawalle-Lange-Haftstrafe-fuer-21-Jaehrigen,gipfeltreffen644.html.

Newell, P. (2019). Trasformismo or Transformation? The Global Political Economy of Energy Transitions. *Review of International Political Economy* 26, no. 1: 25–48.

Nixon, R. (2011). *Slow Violence and the Environmentalism of the Poor*. Cambridge: Harvard University Press.

Öcalan, A. (2013). *Liberating life: Woman's Revolution*. Mesopotamian Publishers.

Oceransky, S. (2011). Fighting the Enclosure of Wind: Indigenous Resistance to the Privatization of the Wind Resource in Soutehrn Mexico. In *Sparking a Worldwide Energy Revolution*, edited by K. Abramsky. Oakland: AK Press: 505–522.

Paoli, G. (2013/2008). *De-Motivational Training*. Berkeley: Cruel Hospice.

Perlman, F. (1985). *The Continuing Appeal of Nationalism*. Detroit: Red & Black.

Perschke, H. (2012). *r-mediabase*. https://www.r-mediabase.eu/index.php?view=detail&id =8624&option=com_joomgallery&Itemid=519.

Pierpont, N. (2009). *Wind Turbine Syndrome: A Report on a Natural Experiment*. Lowell: K-Selected Books.

Porter, G., & Kakabadse, N. K. (2006). Hrm Perspectives on Addiction to Technology and Work. *Journal of Management Development* 25, no. 6: 535–560.

Premalatha, M., Abbasi, T., & Abbasi, S. A. (2014). Wind Energy: Increasing Deployment, Rising Environmental Concerns. *Renewable and Sustainable Energy Reviews* 31: 270–288.

Rahnema, M. (1997). Development and the People's Immune System: The Story of Another Variety of Aids. In *The Post-Development Reader*, edited by M. Rahnema & V. Bawtree. London: Zed Books: 111–129.

Reclus, E. (2013/1905). *Anarchy, Geography, Modernity: Selected Writings of Elisée Reclus*. Oakland: PM Press.

Rivas, S. C. (2015). Consulta Definira Futuro De Inversiones Eolicas: Ocaso O Resplandor. *Noticias*. http://www.noticiasnet.mx/portal/sites/default/files/flipping_book/oax/2015/01 /23/secc_a/files/assets/basic-html/page20.html.

Rodney, W. (1972/2009). *How Europe Underdeveloped Africa*. Washington, DC: Howard University Press.

Romanyshyn, R. (1989). *Technology as Symptom and Dream*. London: Routledge.

Rosenthal, C. (2016). Slavery's Scientific Management: Masters and Managers. In *A New History of American Economic Development*, edited by S. Beckert & S. Rockman. University of Pennsylvania Press.

Rubí, M. (2016). Arranca Segunda Fase De Central Eólica Sureste I. *El Economista*. http:/ /eleconomista.com.mx/estados/2016/03/03/arranca-segunda-fase-central-eolica-sureste-i.

RWE AG. (2012). The Power of Participation. *Essen*. http://www.rwe.com/web/cms/med iablob/en/1716210/data/1701408/6/rwe/responsibility/sustainablecorporate-governance/a cceptance-study/blob.pdf. pp. 19, 6.

RWE Group. (2015). Facts & Figures. *Essen*.

RWE Power. (2016). Das Nachbarschaftsmagazin von RWE Power, 2, 2016. https:// www.rwe.com/web/cms/mediablob/de/3080662/data/496266/2/rwe-power-ag/nachbarsc haft/nachbarschaftsmagazine/hier-rheinisches-braunkohlenrevier/Ausgabe-Indeland-Juni-2016.pdf.

Sale, K. (1991/1985). *Dwellers in the Land: The Bioregional Vision*. University of Georgia Press.

Scott, J. C. (1998). *Seeing Like a State: How Certain Schemes to Improve the Human Condition Have Failed*. Yale University Press.

———. (2017). *Against the Grain: A Deep History of the Earliest States*. Yale University Press.

Selwyn, D. (2020). Martial Mining: Resisting Extractivism and War Together. *London Mining Network*. https://londonminingnetwork.org/wp-content/uploads/2020/04/Martia l-Mining.pdf.

Short, D. (2016). *Redefining Genocide: Settler Colonialism, Social Death and Ecocide*. London: Zed Books.

Silvia, C. R. (2015, Jan 23). Consulta Definira Futuro De Inversiones Eolicas: Ocaso O Resplandor. *Noticias*. http://www.noticiasnet.mx/portal/sites/default/files/flipping_book/oax/2015/01/23/secc_a/files/assets/basic-html/page20.html.

Smil, V. (2016a). *Energy Transitions: Global and National Perspectives*. Santa Barbara: Praeger.

———. (2016b). To Get Wind Power You Need Oil. *IEEE*. http://spectrum.ieee.org/energy/renewables/to-get-wind-power-you-need-oil.

Smith, P. (2014). Soaring Copper Prices Drive Wind Farm Crime. *Wind Power Monthly*. http://www.windpowermonthly.com/article/1281864/soaring-copper-prices-drive-wind-farm-crime.

Sovacool, B. K., Hook, A., Martiskainen, M., Brock, A., & Turnheim, B. (2020). The Decarbonisation Divide: Contextualizing Landscapes of Low-Carbon Exploitation and Toxicity in Africa. *Global Environmental Change* 60: 102028.

Springer, S. (2016). *The Anarchist Roots of Geography: Toward Spatial Emancipation*. Minneapolis: University of Minnesota Press.

Stirling, A. (2015). From Controlling 'the Transition'to Culturing Plural Radical Progress1. In *The Politics of Green Transformations*, edited by I. Scoones, M. Leach & P. Newell. London: Routledge: 54–67.

Sullivan, S. (2010). 'Ecosystem Service Commodities' – a New Imperial Ecology? Implications for Animist Immanent Ecologies, with Deleuze and Guattari. *New Formations: A Journal of Culture/Theory/Politics* 69: 111–128.

———. (2013a). Banking Nature? The Spectacular Financialisation of Environmental Conservation. *Antipode* 45, no. 1: 198–217.

———. (2013b) After the Green Rush? Biodiversity Offsets, Uranium Power and the 'Calculus of Casualties' in Greening Growth. *Human Geography* 6, no. 1: 80–101.

———. (2017). On 'Natural Capital', 'Fairy Tales' and Ideology. *Development and Change* 48, no. 2: 397–423.

Szablowski, D. (2007). *Transnational Law and Local Struggles: Mining, Communities and the World Bank*. Portland: Hart Publishing.

Taussing, M. (1987). *Shamanism, Colonialism and the Wild Man: A Study in Terror and Healing*. Chicago: University of Chicago Press.

Ulloa, A. (2013/2005). *The Ecological Native: Indigenous Peoples' Movements and Eco-Governmentality in Columbia*. London: Routledge.

Veblen, T. (2009/1899). *Theory of the Leisure Class*. New York: Oxford University Press.

Veltmeyer, H. (2013). The Political Economy of Natural Resource Extraction: A New Model or Extractive Imperialism? *Canadian Journal of Development Studies/Revue canadienne d'études du développement* 34, no. 1: 79–95.

Veracini, L. (2014). Understanding Colonialism and Settler Colonialism as Distinct Formations. *Interventions* 16, no. 5: 615–633.

Virilio, P. (1995). *The Art of the Motor*. University of Minnesota Press.

Williams, E. (1944). *Capitalism and Slavery*. University of North Carolina Press.

Winther, T. (2008). *The Impact of Electricity: Development, Desires and Dilemmas*. Berghahn Books.

Wolfe, P. (2006). Settler Colonialism and the Elimination of the Native. *Journal of Genocide Research* 8, no. 4.

Yang, X. J., Lin, A., Li, X. L., Wu, Y., Zhou, W., & Chen, Z. (2013). China's Ion-Adsorption Rare Earth Resources, Mining Consequences and Preservation. *Environmental Development* 8: 131–136.

6

Dismantling the Dam Hierarchies

Jennifer Mateer

The large and imposing infrastructure of dams are often symbols of advancement and ingenuity. These structures are physically large, costly and challenging to achieve in terms of the political manoeuvering, engaging with stakeholders, engineering skills and economic capital necessary for them to be built. As such, dams can provide a lens for understanding the required discourses for legitimizing these projects, the productive power at work that can be linked to values, ethics and epistemologies (Steinberg 1993). This chapter focuses specifically on newly built dams in the Indian state of Himachal Pradesh to discuss the political ecologies of the region from an Anarcha-feminist perspective. The attention of this chapter is particularly important as in Himachal Pradesh, as well as other states and regions, bodies of water are often constructed as an 'other' – and feminized object whose value is based on the utility for humans. This anthropocentric and feminine portrayal of water both add legitimization to the dominations of the 'other'. A re-reading of water and damming, using an Anarcha-feminist Political Ecology (AFPE) would thus be a necessary discursive tool to dismantle discourses that rationalize the destruction of nature and could aid in the dialogue of a humanity that is interconnected and interdependent with nature. To begin the project of this chapter, I will start with a discussion on the rhetorical and cultural significance of dams before discussing an AFPE. I then give an overview of damming discourses in Himachal Pradesh through the lens of AFPE.

In this chapter, I analyse the official literature on dams in Himachal Pradesh since 2010 in combination with first-hand interviews throughout the state. I also examine water conservation campaigns as a juxtaposition between the modernity and abundance discourses that justify the need for dams and the scarcity felt by many people in the state. The analysis is thus a way to identify narratives and ideologies presented by the federal and state governments in order to determine how the management of water adheres to certain dominant ideologies. Understanding these ideologies is important as it is through understanding these discourses, we can present alternatives.

Himachal Pradesh is a state with relative water abundance, located in the western Himalayas. The name of the state means 'in the lap of the Himalayas', and being located in the foothills of the Himalayan mountains gives Himachal Pradesh a geographic advantage with respect to water, as there are high levels of perennial water flowing into the state due to glacier and snowmelt. In fact, glacier run-off is the source of most of the surface water in the state (Bhardwaj 2010). In considering the water running through the state, and through damming structures, my aim is to question the standard logics of water as something to be 'managed, administered, controlled, saved and spent' (Steinberg 1993: 408) and instead consider how power has transformed water into a rhetorical device that highlights the masculine domination of the state over (a feminized) nature.

The abundance of water in the state is socially and culturally reinforced, which is evidenced by children commonly singing folk songs about the ways in which goddesses have brought water to the Himachalese people.[1] Although there is a natural abundance of water, the allocation of resources among citizens of the state is inequitable – making water scarcity a manufactured condition. I do not use the term 'manufactured' as a way to dismiss the water scarcity that people face on a daily basis, but rather to highlight the ways in which the situation has been brought about by inappropriate government policies and management strategies and to dispel the notion of scarcity as natural or as the fault of individual households. Water has been conceptualized through a masculine and economic lens and, as such, has been (neo)liberally manipulated and governed through the creation of dams and the diverging of rivers and streams for economic purposes. In this way, water meets economic needs rather than the needs of the human population and the ecosystem more generally.

Climate change is also challenging water management in Himachal Pradesh. Global climate change is impacting the Himalayan region at a much higher rate than the rest of Asia or the world (Bajracharya 2007; Signh et al. 2011; ICIMOD 2012). For example, there has been a rise in temperatures of 0.6°C each decade in the Himalayan region, whereas the global average is 0.74°C over the last 100 years – so the Himalayan region has experienced in ten years what the rest of the world has experienced (on average) over 100 years (Erkisson et al. 2009). These temperature increases cause stream flows to increase throughout the state as glacier melt occurs at higher than average levels, so in fact, there is *more*[2] water available to citizens, even during seasonal population increases.[3] However, current water management strategies, including damming and run-of-river projects for hydroelectric production, have changed the allocation of water flows, and the ways in various public and private interests intersect with abundance and scarcity.

Due to the political and social controls involved with water allocation, instead of there being an increase of water available to citizens due to climate change, the government of Himachal Pradesh and the federal government are attempting to harness the 'untapped hydropower potential' (Asian Development Bank 2013: 82). Accordingly, the increase in water throughout the state based on glacier melt 'present[s] an opportunity to . . . maximize growth and development' (Hurford et al. 2014: viii). With water as a development asset, the state and the federal governments have begun over 300 hydroelectric projects, including dams and run-of-the-river schemes since 2010[4] (Iyer 2010). This increased the hydropower production by 15,000 MW in 2017 and 23,000 MW by the end of 2020 (Himachal to generate 15,000 MW hydro power 2012). The rationale for the increase in 'hydroelectric projects [is] . . . to meet the energy requirements of Delhi' (Darpan 2007: 591) as well as those of other major cities. These hydroelectric projects have created a surplus of electricity but have also diverted water traditionally supplied to the general public – thereby creating a scarcity that would not otherwise exist. The justification for these projects and new state-based interventions presents a gendered discourse of modernization and domination. Even images (figure 6.1) suggest a masculine and dominating relationship between society and dams.

Figure 6.1 **Image of a muscular man embracing dam infrastructure.** *Source:* Author's own.

CULTURAL SIGNIFICANCE OF DAMS

Although potentially counter-intuitive, the government of Himachal Pradesh, as well as Modi's federal government, has framed the issue of climate change in Himachal Pradesh as an asset. Increased glacier melt has provided an economic opportunity for the state and federal governments, as the abundance of water is now being used to produce energy in new hydropower developments. The economic performance of the state has begun to rise due to the way in which the increased flow of water in the state is being used. Although not released by the state, there are images and advertisements that are indicative of the current political rationalities. The images used to frame this infrastructural development in the state consist of a muscular man embracing the walls of a dyke (figure 6.1). This is indicative of the economic and gendered rationality in Himachal Pradesh that the natural environment is feminized and in need of control through the masculine discourses and actions of resource management.

As suggested within the literatures around the benefits of the dam, as well as the image presented figure 6.1, in Himachal Pradesh (and India more generally) dams have become symbols of triumph over nature, with phrases calling the dams feats of civil engineering and the conquest of the natural environment (Kaminsky and Long 2011; Goyal 2015; India Today 2018; Central Electricity Authority 2018; Energy World 2018 Hurford et al. 2014). The dams are therefore more than just energy-producing giants; they are also 'concrete, rock and earth expressions of the dominant ideology of the technical age: icons of economic development and scientific progress' (McCully 1996: 2–3). The building of dams in the state has nonetheless been controversial, particularly due to the ecological impacts felt by many small communities in the state. As previously stated, the rationale for the increase in 'hydroelectric projects [is]. . . . to meet the energy requirements of Delhi' (Darpan 2007: 591) and as such many local residents described feeling resentful of these plans, because they 'are the ones suffering the [ecological] consequences so others can benefit' (Amita[5] Interview, Manali). However, the dissent towards the number of hydroelectric projects has not meant a halting or even slowing of their build. Instead, the rationale for these dams is part of a Modi-led government's regime of truth (Foucault 1980) regarding nature and progress, and thus halting development would be quite literally unthinkable (Foucault 1970: xv). The construction of dams is 'common sense', further justifying industrialization (Rogers and Schutten 2004).

The discourses of progress and modernization shape what interventions are possible by the state and general populace because what is considered 'common sense' is a powerful method of control (Gramsci 1971; Williams 1977). Thus, the meanings people assign to the new dams of Himachal Pradesh demonstrate key ideologies surrounding the environment, economic growth and progress. Exploring these ideologies is important for understanding state-based interventions in environmental management and resource development.

ANARCHA-FEMINISM AND POLITICAL ECOLOGY

Anarchism has always been feminist; however, 'history tells us that being an anarchist doesn't make one automatically non-sexist' (Sethi 2020: n.p.). Similarly, being a feminist does not necessitate that one is against all forms of domination (ibid.). This is where Anarcha-feminism becomes an important bridge. In the 1860s, the anarchist critique of capitalism was combined with a critique of patriarchy. Bakunin claimed a combined rejection of capitalism and patriarchy was necessary given the way laws and government 'subject women to the absolute domination of the man . . . [thus] equal rights must belong to men and women' (Bakunin 1971: 396) in order for there to be true freedom and independence. However, the specifics of Anarcha-feminism began with authors Emma Goldman (Goldman 2006; 2012; Falk 2019), Voltairine de Cleyre (De Cleyre and Havel: 914; Campbell 2013; Avrich 2018) and Lucy Parsons (Parsons 1890; Ashbauugh 2013; Jones 2017) in the late nineteenth and early twentieth centuries. As a political ideology, these female Anarchists saw their goal as not only abolishing the capitalist state but also ending the structural violence of patriarchal domination and gendered oppression. These goals are considered inseparable because 'the struggle against patriarchy is an inherent part of the struggle to abolish the state and abolish capitalism, as they believe that the state itself is a patriarchal structure' (Hofmann-Kuroda 2017: 1). Thus, the emancipation of women would lead to a faster transition away from capitalism and dominant state-based governing structures.

The Siren newsletter, first published in 1971, was where Anarcha-feminist theory and debate first emerged and where a manifesto was originally published[6] (Ehrlich 1979; Kurin 2004). Many of the discussions in these early writings were focused on the need to 'outgrow' conventional state structures and management techniques rather than overthrowing the state. These feminist theorists further worked to develop an Anarcha-feminist ethic of care, developing an understanding of the way in which relationships beyond oppression were part of a feminine ethic (Weber 2009; Falk 2019; Goldman 2012).

Similar to Anarcha-feminism, Feminist Political Ecology (FPE) examines everyday resource-based practices that produce spatial differences and gendered inequalities (Harris 2009; Laurie 2005; O'Reilly 2010; Sultana 2009; Zwarteveen and Meinzen-Dick 2001). Further, political ecology connects the ways in which nature is regarded and managed is implicated in how society is organized. In South Asia, FPE scholarship on water highlights the various ways in which gender is experienced, reinforced and agitated within communities and households based on the different experiences of resource practices, access, control and governance. In this way, the authors contributing to FPE literature highlight the importance of socially produced actions in the shaping of gender roles and ideologies within the complexity of water governance and management. One example of this is a study by Sultana (2009) which discusses, in part, the decision-making processes of women collecting water in Bangladesh. Due to the arsenic poisoning in many of the local tube wells, some women must decide whether to collect water from women-dominated spaces with tainted water or venture into male-dominated spaces with untainted water. Both spaces, whether male or female-dominated, are both unsafe and safe in different ways, either through arsenic contamination, or the potential for physical and mental abuse. In this way, collecting water is an embodied task that produces specific gender identities. As such, Sultana's study exemplifies the ways in which power and gender coalesce with everyday practices of resource management and control.

FPE further 'seeks to understand and interpret local experience in the context of global processes of environmental and economic change' (Rocheleau et al. 1996: 4), which connects with the Anarcha-feminist approach of connecting patriarchy with capitalism and the state. By using both Anarcha-feminism and FPE, scholars can better understand the lived experiences of individuals marginalized by the structural violence of patriarchy, tracing these power relations from the local to the global scales. This approach is particularly important since access to water, like many resources, has been and continues to be dependent on gender, race, social capital and economic capital, similar to what has been described within the large-scale dam and irrigation projects (Braun 2011; Harris 2008). AFPE then can be the 'imaginative thinking and practice to rescue nature from corporate control, financialization, and the proprietary exploitations of biogenetic capitalism' (Tarleton 2014: n.p.).

DISCOURSES OF DAMMING

The development of an AFPE as a critical frame was motivated by my close reading of the project documents, media releases and government literature regarding the development of hydroelectric projects in Himachal Pradesh. The themes most strongly presented in these texts included nature as feminized, the necessity of economic productivity and technology as powerful and modern. The entanglements of gender, nature, capitalism and modernization are the core issues of AFPE. These are important relationships, which if we recognize as contextually interconnected could lead to the 'out growing' project centred in Anarcha-feminism.

Nature as Dangerous and Feminine

In government documents, as well as water conservation campaigns within the state, readers and citizens are consistently made aware of the centrality of water to life in Himachal Pradesh and human life more generally. Images of clear and clean water, seen only in abundance, are used. Images like clear and blue water watering plants and agriculture, splashing over small children are being used in a river-run project – none of it in its natural 'unused' state. The 'web of life', which is predicated on waters cleanliness and availability, is seemingly centred around humans, with plants and animals presented only in their relation to human use. Rivers and lakes are considered dirty due to the level of silt, as well as wild and unpredictable (Power Technology 2012). For the dams in Himachal Pradesh, the chosen metaphors for discussing water in the region include lots of mention of fluidity, which has been linked to femininity, chaos and emotion (Hayles 1992; Irigaray 1985), whereas solidity is linked with order and masculinity (Rogers and Schutten 2004). It, therefore, makes sense that female pronouns are used in relation to the water flowing through the state, which reinforces the gendered dualisms that dominate many Western philosophical and religious traditions. Women stand in opposition to civilization, as they are more natural, nurturing but also moody and erratic (Merchant 1980; Schott 1988). Vandana Shiva describes this influence of Western thought in India as being one of the many consequences of colonialism (1988). Under British rule in the mid-eighteenth century, India was forced to move away from a cosmological view of nature Prakriti, which valued a reciprocal and living relationship between humans and the ecosystem and instead adopted the notion of humans needing to dominate nature (ibid.). Shiva suggests that this not only caused a devaluation of the natural environment and more-than-human beings but also 'the death of the feminine principle' (Shiva 1988: 42). The ontology of the feminine principle is founded in a continuity between humans and nature, rather than the dualism common in Western thought. Shiva claims that this way of being in the world common in India provided an ethical context that made exploitation and domination of environmental resources unthinkable (1988: 41).

There is also gendered language used to describe the usage of water within the government literatures and conservation campaigns. As previously mentioned, the increased number of hydroelectric campaigns have produced a state surplus of electricity but has also diverted water traditionally supplied to the general public – thereby creating a manufactured scarcity that would not otherwise exist. In order to manage this scarcity and maintain a surplus of electricity, water conservation campaigns have been initiated by the state via the Central Water Commission, and by various NGOs such as The Global Water Partnership, the Himalayan Nature & Environment Preservation Society and the Himalayan Eco Horticulture Society. These conservation strategies include rainwater harvesting techniques and priced metering to encourage conservation (Government of Himachal Pradesh 2014). Some of the signage and slogans which promote these modes of conservation are as follows: Water Saved is Energy Generated; Save Water, Save Life & Save the World (see figure 6.2); Water is Precious! Be Wise in Using It!; Don't Waste Water or the World Won't Get Any Better; and See Mother Nature's Face Turn Sour When You Take a Long-Long Shower (Department of Irrigation and Public Health n.d.). These campaign slogans seem to indicate the scarcity and capricious nature of water – the only hope for reliable supply being self-regulation and the technological feats of engineering provided through dams and run-of-river projects.

The Modernization Project

As many authors have noted, water flows have been progressively 'modernized' with technology and have become more connected to capital accumulation strategies in an evolution towards commodification (Bakker 2005; Braun 2011; Castro 2013;

Figure 6.2   Sign saying Save Water, Save Life & Save the World. *Source:* Author's own.

Loftus 2006; Shiva 2002; Sultana 2013; O'Reilly 2006). Part of this shift results from development strategies that advocate for a specific version of 'progress'. As Vandana Shiva says, 'A sustainable and clean river is not a productive resource in this view: it needs to be developed with dams in order to become so' (1989: 82). As such, the waters of Himachal Pradesh are useless unless developed by the state as part of a 'modern' domestic and international economy. Alternatively, once developed, 'Modi said these hydro projects will bring prosperity to the State and other parts of the nation' (Press Trust of India 2016, October 19).

As the discourses within the government literature and news sources suggest, technological modernization within the waterscape has become a key rationale for the increase in hydroelectric power production throughout Himachal Pradesh. There is even an element of triumphalism in these documents when the dams are discussed as feats of technology and engineering. For example, Rampur in Himachal Pradesh hosts one of the world's biggest hydroelectric plants and the biggest plant in whole India (Directorate of Energy 2018; Joshi 2007; Central Water Commission 2018;

Central Water Commission 2017; The World Bank 2007). The plant, which was approved and funded by The World Bank, has six vertical axis Francis turbine units, double-circuit lines and a concrete-lined underground head-race tunnel to steel penstocks – all marvels of the modern era (Himachal Directorate of Energy 2018; Central Water Commission 2018; Central Water Commission 2017). Subsequently, the ways in which the provision of water is viewed – the dimensions emphasized – have also shifted from water as a resource that satisfies basic needs to now something that will bring 'development' and/or 'modernization' (Walsh 2011). The water in Himachal Pradesh, having been dominated by human engineering, has now conformed to progress and the modernization project.

The discourses of masculinity and modernization become even more clear when examining those bodies of water that have yet to be 'developed'. For example, one damming project in the state 'has only harness[ed] 10351 MW capacity of total approximate of 27436 MW' (Balwant 2018, [*sic*]). This project, when compared to the possibilities brought through modernization, development and progress, is lacking – the resource wasted. To ensure this modernization stays on track, India's Parliamentary Standing Committee on Energy has changed energy policy so that *any* hydroelectric project that is proposed or upgraded above 25 MW would also be considered as 'renewable', and thus would eligible for further subsidies and other government funding (India's Parliamentary Standing Committee on Energy 2017).

Dam Hierarchies

In Himachal Pradesh, and more generally throughout the world, the building of dams is considered an immense achievement, due primarily to the discourses of progress, modernization and development that accompany the building of hydroelectric facilities (Rogers and Schutten 2004; Kaika 2006; Hurford et al. 2014; Kaminsky 2011). This type of logic presents a rhetoric of dominance of the natural environment – that humans become 'natures master' (Althusser 1971 in Rogers and Schutten 2004). 'There is a conquering at work here, in which triumph comes at the cost of another's subordination' (Rogers and Schutten 2004: 274), the subordination of a feminized water. This way of thinking about the relationship between humans and nature has been common in Western traditions for centuries (Rifkin 1987), a tradition that was brought to India during colonialism. The progress experienced in Himachal Pradesh has altered many of the waterways, and this trend is only bound to increase with Modi's use of resource management for economic growth (Ruparelia 2015; Narlikar 2017; Schöttli and Paul 2016).

The economic growth and development of the hydroelectric capacity in Himachal Pradesh require both the alienation from and domination over the bodies of water in the region. This is how the dams and run-of-river projects are justified, and the ecological consequences overlooked. An AFPE sees the alienation of humans from nature as a prerequisite to rationalize environmental degradation. It is therefore necessary within AFPE to *outgrow* anthropocentric thinking to ensure humans maintain an interconnected and relational connection with nature.

Water management generally requires the power to control and alter the way water flows. Power is not simply part of the building of hydroelectric facilities but also through the discourses of progress and modernization that make the alienation and domination of water 'natural'. AFPE works to undo the ways in which nature, and in this case bodies of water, have been transformed into the 'other' – either a lifeless or hostile resource. Globally we have outgrown the Western concept of masculine development and progress that has dominated resource management and gender relations – a way of thinking that allowed for the domination of entities and objects deemed feminine. However, with AFPE, nature is no longer the 'other', and thus objectification is no longer justifiable. Dams can no longer be symbols of human domination and progress. Through this process, AFPE ensures all people work to reconceptualize themselves in relation to the non-human world beyond the hierarchy currently imposed. In this way, gender equality and environmental degradation are both addressed – standing for egalitarian, non-hierarchical systems and avoiding the 'othering' of feminized entities.

CONCLUSION: OUTGROWING CAPITALISM AND PATRIARCHY

This reading of the discourses surrounding the building of hydroelectric projects in Himachal Pradesh has confirmed some previously understood conceptions but presents new opportunities for engaging with an AFPE. Despite the various ways in which the high number of dams and run-of-river projects have altered the ecosystem throughout the state, the discourses presented show very few linkages to environmental stewardship. Instead, the industry is painted as part of the necessary progress in India's growing economic development. This progress is naturalized through the human/nature dualism present that advocates for human dominion over nature – an ideology that paints domination as an accomplishment of human ingenuity in engineering and technologies. Woven throughout these discourses is a feminization of water and nature more generally. This is a common theme, however, it is still a relatively new logic to India,[7] which has a history of understanding of ecosystems through the feminine principle of the cosmos – the Prakriti, which describes the entanglement of humans and nature, rather than seeing the relationship as a dichotomous. An AFPE also confronts this dichotomy, with the added analysis of the domination of those entities deemed feminine, thereby critiquing capitalism and patriarchy.

As a political ideology, the goal of AFPE is the abolishment of the capitalist state, ending the structural violence of patriarchal domination and gendered oppression, and critiquing the use of nature for capital accumulation and political means. These goals are inseparable because, as previously mentioned, 'the struggle against patriarchy is an inherent part of the struggle to abolish the state and abolish capitalism' (Hofmann-Kuroda 2017: 1). This too would remove the natural environment from extractivism and resource management for the purpose of capital accumulation.

There are instances where managing water resources can be a method of mitigating scarcity or improving conservation. In theory, these are positive strategies, however in Himachal Pradesh, water management, and even sustainability, is being used as a rhetorical tool to control the way water is used by citizens in order to ensure continued economic growth and development via energy production. Specifically, by ensuring that the general population conserves water, and by limiting the access to water by citizens, hydropower production is maintained and continues to grow. As such, the framing of these dams as necessary for conservation and environmental sustainability serves a political agenda, denying a plural reading of the situation (Fairhead and Leach 1996), disciplining populations for political and economic purposes, while also limiting the ways in which scarcity and sustainability can be thought of and addressed. Protecting the environment and sustainability is not a trend; it is indeed essential for the survival of many beings on this little blue marble.

To conclude this chapter, I would like to re-examine figure 6.1. Today, the governing mentality is one that privileges economic development and alienates people from the natural environment. Alienation, in this case, has encouraged a gendered reading of the relationship between humans and water and is exemplified in the image of a man – muscles budging as his arms attempt to grip a dam structure. This image is indicative of a common sense rationality – the valiant masculine domination of (feminized) water, an erratic and underutilized resource, is in need of control. While this rationality follows a patriarchal and capitalist logic, what if we were to see an image of the relationship between humans and water and the building of dams from the logic of AFPE? This image would communicate a reciprocal relationship between humans and nature. Perhaps it would

Figure 6.3 Women in Vanuatu performing the Ëtëtung. *Source:* Taken by Aude Emilie-Dorion (2019).

look a bit like figure 6.3, which illustrates the relationship with water on a small island in Vanuatu (Emilie-Dorion 2019). Here, women practice water music (The Ётětung) to celebrate the connection that exists between people and the sea (ibid.). In comparison to figure 6.1, figure 6.3 presents us with an AFPE perspective and rationality.

Like figure 6.3, AFPE can address the intense power inequalities of global society (including the human and more-than-human communities) along class, gender and racialized lines, and is able to build on the limited success of state action by eliminating the hierarchical structures in resource and environmental management. For example, local organizing, citizen ownership of limited power structures and community empowerment are highly sustainable, peaceful and positive forces.

Resiliency, solidarity and redistributive actions are the best way to respond to the ongoing environmental degradation and crises. These processes are outside of any capitalist production model or economic development, so our response to environmental issues should be thought of as growing – growing in resilience and the capacities of local organizations. Individual action, corporate responsibility or the extensions of state power are not the way forward. As Anarcha-feminists suggest, it's time to outgrow these notions and reconnect with the reciprocal relationships previously experienced.

NOTES

1. One of the most popular songs celebrating water is 'Panni Peena Ho', which translates to 'Water to Drink'. I met a group of school children, who after hearing I was doing research on water issues sung the song for me and even performed a short dance.

2. This will eventually be followed by decreased river flows as the glaciers recede, but while the glaciers are in a period of increased melting it does mean more water in the present (Eriksson et al. 2009). Although this could lead to further conservation rationale, at this time the increase due to climate change has not been addressed by the government in this way. Instead, the increase has only been considered an asset for economic development through power production.

3. The competition for water increases seasonally with an influx of tourist and service populations (Ridolfi 2014). Peak tourist season is generally between June and August during India's hottest months. Visitors during these months, predominantly Indian nationals, come to the mountainous region because the temperatures are much lower than in the more southern states of the country (Jreat 2004).

4. Interestingly, the hydroelectric projects – and thus control – are mostly in the hands of the federal government, rather than the state of Himachal Pradesh (Koschel 2013). In fact, 'the Sates function under the guidance of the Central government' (Saxena and Kumar 2010: 1), so even when there is a state-based hydropower project in place there is still a significant amount of control held by the federal government. As such is comes as no surprise that often these projects may economically benefit India, local populations are neglected to face the consequences of development with limited support.

5. This participant in particular raised concerns about stray animals that depend on reliable watering holes and ponds. These watering sources have been modified with changes due

to hydropower diversions, so places where animals originally found water have sometimes no longer exist.

6. 'Who We Are: The Anarcho-Feminist Manifesto' was published in Siren's first issue.

7. In the Siva-Shakti tradition of India, the male deity, Siva is passive whereas Shakti, the feminine, is active. As Vandana Shiva describes, 'Without Shakti, [Siva], the symbol for the force of creation and destruction, is as powerless as a corpse. The quiescent aspect of [Siva] is, by definition, inert' (1988: 39).

REFERENCES

Ashbaugh, C. (2013). *Lucy Parsons: An American Revolutionary*. Haymarket Books.

Asian Development Bank. (2010). *Climate Change Adaptation in Himachal Pradesh: Sustainable Strategies for Water Resources*. Manila: Asian Development Bank.

Avrich, P. (2018). *An American Anarchist: The Life of Voltairine de Cleyre*. AK Press.

Bajracharya, S. R., Mool, P. K., and Shrestha, B. R. (2007). *Impact of Climate Change on Himalayan Glaciers and Glacial Lakes: Case Studies on GLOF and Associated Hazards in Nepal and Bhutan*. Kathmandu: International Centre for Integrated Mountain Development.

Bakker, K. (2005). Neoliberalizing Nature? Market Environmentalism in Water Supply in England and Wales. *Annals of the Association of American Geographers*, 95.3: 542–565.

Bakunin, M. (1971). *Bakunin on Anarchy*, edited by Sam Dolgoff. New York: Vintage Books.

Balwant, S. (2018, Jan 28). *HP GK – Hydroelectric Power Project in Himachal Pradesh*. District Mandi. https://disttmandi.com/hp-gk-hydroelectric-power-project-himachal-pradesh/.

Bhardwaj, A. (2014). *Directory of Water Resources in Himachal Pradesh*. State Centre on Climate Chande: Himachal Pradesh State Council for Science Technology and Environment. http://www.hpccc.gov.in/documents/Water%20Sources%20in%20Himachal%20Pradesh .pdf.

Braun, Y. A. (2011). The Reproduction of Inequality: Race, Class, Gender, and the Social Organization of Work at Sites of Large-Scale Development Projects. *Social Problems*, 58.2: 281–303.

Campbell, M. (2013). Voltairine de Cleyre and the Anarchist Canon. *Anarchist Developments in Cultural Studies*, 2013.1: 64–81.

Castro, J. E. (2013). Water is not (yet) a Commodity: Commodification and Rationalization Revisited. Human Figurations. *Long-term Perspectives on the Human Condition*, 2.1. http:/ /hdl.handle.net/2027/spo.11217607.0002.103.

Central Electricity Authority. (2018). *Hydro Power Development in Himachal Pradesh*. Delhi: Government of India. http://cea.nic.in/reports/monthly/hydro/2018/state_power-02.pdf.

Central Water Commission. 2017. *Annual Report*. Shimla: Ministry of Water Resources. http: //www.cwc.gov.in/sites/default/files/CWC_AY_2016-17.pdf.

———. 2018. *Hydrological Observation on Various Rivers in Himachal Pradesh*. TCIL Reference No. CWC-2018-TN000168. Shimla: Government of India. http://cwc.gov.in/sites/ default/files/supportservice2018.pdf.

Darpan, P. 2007. Regional News: Himachal Pradesh. *2nd Annual Current Affairs Special*. http://www.pdgroup.upkar.in/.

De Cleyre, V., and Havel, H. (1914). *Selected Works of Voltairine de Cleyre*. New York: Mother Earth Publishing Association.

Department of Economics and Statistics. (2018) *Economic Survey*. Shimla: HP Economics & Statistics Department. http://himachalservices.nic.in/economics/ecosurvey/en/power .html.

Department of Irrigation and Public Health. (n.d.). *Government of Himachal Pradesh*. http:// www.hpiph.org/.

Directorate of Energy. (2018). *Disaster Management Plan*. Shimla: Government of Himachal Pradesh.

Ehrlich, H. J. (Ed.). (1979). *Reinventing anarchy: What Are Anarchists Thinking These Days?* London: Routledge.

Emilie-Dorion, A. (2013, 7 June). In Vanuatu, Women Draw Strength from the Rhythm of the Ocean. *Earth Journalism Network*. https://earthjournalism.net/stories/in-vanuatu-wom en-draw-strength-from-the-rhythm-of-the-ocean.

Energy World. (2018). Himachal Pradesh Approves Amendments in Hydro-Policy to Promote Investment. *The Economic Times*. https://energy.economictimes.indiatimes.com/news/power/ himachal-pradesh-approves-amendments-in-hydro-policy-to-promote-investment/64092306.

Eriksson, M., Jianchu, X., Shrestha, A. B., Vaidya, R. A., Nepal, S., and Sandström, K. (2009). *The Changing Himalayas: Impact of Climate Change on Water Resources and Liveli-hoods in the Greater Himalayas*. Kathmandu, Nepal: International Centre for Integrated Mountain Development (ICIMOD). ISBN 9789291151141. http://www.preventionweb. net/publications/view/11621.

Erratic Water Supply Gives Residents Sleepless Nights. (2014, December 26). *The Times of India*. http://timesofindia.indiatimes.com/city/aurangabad/Erratic-water-supply-gives-re sidents-sleepless-nights/articleshow/45645045.cms.

Fairhead, J., and Leach, M. (1996). *Misreading the African Landscape: Society and Ecology in a Forest-Savanna Mosaic* (Vol. 90). Cambridge: Cambridge University Press.

Falk, C. (2019). *Love, Anarchy, & Emma Goldman: A Biography*. Rutgers University Press.

Foucault, M. (1970). *The Order of Things: An Archeology of the Human Sciences*. New York: Vintage.

———. (1980). *Power/Knowledge: Selected Interviews and Other Writings 1972–1978*. C. Gordon (Ed.). New York: Pantheon Books.

Goldman, E. (2006). *Vision on Fire: Emma Goldman on the Spanish Revolution*. AK Press.

———. (2012). *Quiet Rumours: An Anarcha-Feminist Reader*. AK Press.

Goyal, A. (2015). *RBS Visitors Guide India – Himachal Pradesh: Himachal Travel Guide*. Jaipur: Data and Expo India Pvt.

Gramsci, A. (1971). *Selections from the Prison Notebooks of Antonio Gramsci*. New York: International Publishers.

Harris, L.M. (2008). Water Rich, Resource Poor: Intersections of Gender, Poverty and Vul-nerability in Newly Irrigated Areas of Southeastern Turkey. *World Development* 36, no. 12: 2643–2662.

———. (2009). Gender and Emergent Water Governance: Comparative Overview of Neo-liberalized Natures and Gender Dimensions of Privatization, Devolution and Marketiza-tion. *Gender, Place and Culture* 16, no. 4: 387–408.

Hayles, N. K. (1992). Gender Encoding in Fluid Mechanics: Masculine Channels and Femi-nine Flows. *Differences* 4, no. 2: 16–44.

Himachal to Generate 15000 MW Hydro Power by 2017: CM. (2012, March 21). *The Economic Times*. http://economictimes.indiatimes.com/industry/energy/power/himachal -to-generate-15000-mw-hydro-power-by-2017-cm/articleshow/12485483.cms.

Hofmann-Kuroda, L. (2017). Anarcha-Feminists of the Past and Present Are an Inspiration for Today. *Wear Your Voice Magazine*. https://wearyourvoicemag.com/identities/feminism /anarcha-feminists-past-present.

Hurford, A. P., Wade, S. D., and Winpenny, J. (2014). *Himachal Pradesh, India Case Study: Harnessing Hydropower*. London: Department for International Development, Government of the United Kingdom. http://dx.doi.org/10.12774/eod_cr.august2014.hurfordaet al01.

ICIMOD. (2012). *North India, Himalayas to be Worst Hit by Climate Change*. http://www .icimod.org/?q=8800.

India Today. (2018, October 22). *Bhakra-Nangal Dam: History, Features and Facts About the Second Tallest Dam in Asia*. https://www.indiatoday.in/education-today/gk-current-affairs /story/bhakra-nangal-dam-things-you-should-know-about-the-second-tallest-dam-in-asia-1372739-2018-10-22.

Irigaray, L. (1985). *This Sex Which Is Not One*. Ithaca, NY: Cornell University Press.

Iyer, R. (2010, March 17). 294 Hydropower Projects Being Built in Himachal Pradesh. *Top News*. http://www.topnews.in/law/294-hydropower-projects-being-built-himachal-pradesh -212241.

Jones, J. (2017). *Goddess of Anarchy: The Life and Times of Lucy Parsons, American Radical*. New York: Basic Books.

Joshi, H. (2007). *Rampur Hydropower Project: International Rivers*. https://www.international rivers.org/resources/rampur-hydropower-project-3655.

Jreat, M. (2004). *Tourism in Himachal Pradesh*. New Delhi: Indus Publishing.

Kaika, M. (2006). Dams as Symbols of Modernization: The Urbanization of Nature Between Geographical Imagination and Materiality. *Annals of the Association of American Geographers* 96, no. 2: 276–301.

Kaminsky, A. P., and Long, R. D. (2011). *India Today: An Encyclopedia of Life in the Republic* (Vol. 1). Abc-Clio.

Koschel, H. (2013). *Energy and Employment: Case Study Hydropower in India* (KFW Position Paper). Cologne: KFW. https://www.kfw-entwicklungsbank.de/Download-Center/PDF -Dokumente-Positionspapiere/2013_01_Energie-und-Besch%C3%A4ftigung_EN.pdf.

Kurin, K. (2004). Anarcha-Feminism: Why the Hyphen? In *Only a Beginning: An Anarchist Anthology*, edited by A. Antliff. Vancouver: Arsenal Pulp Press: 257–263.

Laurie, N. (2005). Establishing Development Orthodoxy: Negotiating Masculinities in the Water Sector. *Development and Change* 36, no. 3: 527–549.

Loftus, A. (2006). Reification and the Dictatorship of the Water Meter. *Antipode* 38, no. 5: 1023–1045.

McCully, P. (1996). *Silenced Rivers: The Ecology and Politics of Large Dams*. London: Zed Books.

Merchant, C. (1980). *The Death of Nature: Women, Ecology and the Scientific Revolution*. San Francisco: Harper & Row.

Narlikar, A. (2017). India's Role in Global Governance: A Modification? *International Affairs* 93, no. 1, 93–111.

O'Reilly, K. (2006). 'Traditional' Women, "Modern' Water: Linking Gender and Commodification in Rajasthan, India. *Geoforum* 37, no. 6: 958–972.

———. (2010). Combining Sanitation and Women's Participation in Water Supply: An Example from Rajasthan. *Development in Practice* 20, no. 1: 45–56.

Parsons, L. E. (2004). *Lucy Parsons: Freedom, Equality & Solidarity – Writings & Speeches, 1878–1938*, edited by G. Ahrens and R. Dunbar-Ortiz. Chicago: Charles H. Kerr.

Power Technology. (2012). *Rampur Hydro Electric Power Project, Himachal Pradesh*. https://www.power-technology.com/projects/rampur-hydro-electric-power-project-himachal-pradesh/.

Press Trust of India. (2016, October 19). Modi Launches 3 Hydroelectric Projects of 1732 MW at Himachal Pradesh. *YourStory*. https://yourstory.com/2016/10/modi-hydro-electric-projects-hp.

Rifkin, J. (1987). *Time Wars*. New York: Simon & Schuster.

Rocheleau, D., Thomas-Slayter, B., and Wangari, E. (Eds.). (1996). *Feminist Political Ecology: Global Issues and Local Experiences*. London: Routledge.

Rogers, R. A., and Schutten, J. K. (2004). The Gender of Water and the Pleasure of Alienation: A Critical Analysis of Visiting Hoover Dam. *The Communication Review* 7, no. 3: 259–283.

Ruparelia, S. (2015). 'Minimum Government, Maximum Governance': The Restructuring of Power in Modi's India. South Asia. *Journal of South Asian Studies* 38, no. 4: 755–775.

Sano, Y. (2013). *Its Time to Stop This Madness – Philippines Plea at UN Climate Talks*. Responding to Climate Change. https://www.rtcc.org/2013/11/11/its-time-to-stop-this-madness-philippines-plea-at-un-climate-talks/.

Saxena, P., and Kumar, A. (2010). Hydropower Development in India. *Proceedings of the International Conference on Hydraulic Efficiency Measurement*, Roorkee, India. International Group for Hydraulic Efficiency Measurement. http://www.ighem.org/meetings.html.

Schott, R. M. (1988). *Cognition and Eros: A Critique of the Kantian Paradigm*. Boston: Beacon Press.

Schöttli, J., and Pauli, M. (2016). Modi-Nomics and the Politics of Institutional Change in the Indian Economy. *Journal of Asian Public Policy* 9, no. 2: 154–169.

Sethi, S. (2002). Anarcha Feminism: The Beginning of the End of All Forms of Oppression. *Feminist India*. https://feminisminindia.com/2020/02/03/anarcha-feminism-beginning-end-forms-oppression/.

Sharma, V. (2016, July 13). NABARD Takes Baby Steps to Revive Water Table in Punjab, Haryana. *Economic Times*. http://economictimes.indiatimes.com/articleshow/53194623.cms?utm_source=contentofinterest&utm_medium=text&utm_campaign=cppst.

Shiva, V. (1988). *Staying Alive: Women, Ecology and Survival in India*. Kali for Women.

———. (1989). Development, Ecology, and Women. In *Healing the Wounds: The Promise of Ecofeminism*, edited by J. Plant: 80–90.

———. (2002). *Water Wars: Privatization, Pollution and Profit*. Cambridge: South End Press.

Steinburg, T. (1993). 'That World's Fair Feeling': Control of Water in 20th-Century America. *Technology & Culture* 34: 401–409.

Sultana, F. (2009). Fluid Lives: Subjectivities, Gender and Water in Rural Bangladesh. *Gender, Place and Culture* 16, no. 4: 427–444.

———. (2013). Water, Technology, and Development: Transformations of Development
Tarleton, J. (2014, September 12). Interview: Naomi Klein Breaks a Taboo. *Indypendent*. https://indypendent.org/2014/09/12/interview-naomi-klein-breaks-taboo.

Technonatures in Changing Waterscapes. *Environment and Planning D: Society and Space* 31, no. 2: 337–353.

The World Bank. (2007). *Rampur Hydropower Project*. https://www.worldbank.org/en/news/feature/2007/09/13/rampur-hydropower-project.

Walsh, A. (2011). The Commodification of the Public Service of Water: A Normative Perspective. *Public Reason* 3, no. 2: 90–106.

Weber, L.G. (2009). *On the Edge of All Dichotomies: Anarch@-Feminist Thought, Process and Action, 1970–1983*. Dissertation. Department of History, Wesleyan University.

Williams, R. (1977). *Marxism and Literature*. New York: Oxford University Press.

Zwarteveen, M. Z., and Meinzen-Dick, R. (2001). Gender and Property Rights in the Commons: Examples of Water Rights in South Asia. *Agriculture and Human Values* 18, no. 1: 11–25.

7

The Conservation of Anarchy

Ethnographic Reflections on Forest Policies and Resource Use

Philipp Zehmisch

The conservation paradigm challenges the still persistent anthropocentric discourse according to which 'modern man' is vested with the right to subdue planet Earth and its environment. In this hegemonic discourse of 'modernity', the state assumes the role of an agent of 'progress' exploiting available natural resources in order to achieve 'development'. In contrast, conservation policies broadly aim to preserve the environment instead of simply extracting, selling, consuming, and destroying it. While subscribing to the conservationist agenda, however, many actors in this field still adhere to an epistemological bias that is characteristic of 'modernity' (cf. Latour 1993): opposing 'culture' to 'nature', humans continue to be juxtaposed to the environment instead of acknowledging the interdependence and contextual inseparability of human and more-than-human actors and spheres.

Accordingly, a 'pure' environment can only exist without 'humans' living in it; expressed in a drastic and sarcastic way – which is sadly enough not entirely removed from reality – this implies, for example, that visitors to national parks desire to watch wildlife in its 'nature' without those human inhabitants who had cohabited with the flora and fauna of these parks since time immemorial. Another paradoxical tragedy of this paradigm shift is that those 'indigenous'[1] peoples, whose marginalization, discrimination, displacement, exploitation and extinction has since several centuries been justified by constructing them as 'primitive' and 'children of nature', and thus as obstacles to 'modernity' and 'progress', are now increasingly displaced by people desiring to conserve 'nature'.

Characteristic of this conundrum, contemporary conservation policies target and affect politically autonomous and self-sustaining communities inhabiting remote, biodiverse zones of the planet. Especially in the Global South, the demand to protect the vulnerable environment is often weighed against the land rights of 'indigenous' hunting and gathering, herding or peasant peoples who have, for ages, survived by utilizing available resources without usually depleting them. These communities have often evaded state influence by following anarchic strategies in order to maintain their, often partial, autonomy and autarky from the state and markets (cf. Graeber 2008; Scott 2009). Conservation policies are thus liable to function as direct or indirect attack on the subsistence and self-governing practices of anarchic communities.

This contribution critically discusses the impact of conservation efforts on anarchic communities living in biodiverse, marginal forest spaces. It aims to ask for the potential of contemporary conservation policies to impede or, on the contrary, even to enable anarchy by providing a legal framework that works to protect a community's independence of the state and the market. Seeking to answer this 'dialectic' between conservation and anarchy, I will concentrate on a distinct ethnographic setting: the Andaman Islands in the Bay of Bengal, where I conducted 22 months of fieldwork between 2006 and 2016.[2] Here, two anarchic communities – one indigenous, the other composed of 'indigenous' migrants – have been affected by discourses and practices of the state in very different ways; while both share similar ecological spaces, they are confronted with diverging historic, socio-economic and political conditions defining their spectres of autonomy and sustenance.

The first example highlights the original inhabitants of the tropical rainforests of the Andamans. These indigenous hunter-gatherers have been severely decimated by colonization – by the British Empire (1858–1942), Japan (1942–1945) and India (since 1947) – and frequent extensions of the 'frontier' into the forests that have been accompanied by a gradual ethnocide (Venkateswar 2004). However, some remaining indigenous communities continue to live an anarchic way of life until today, and confirm the subject position of the modern state by epitomizing an 'uncivilized, exotic other'. Those indigenes are at the centre of an ongoing debate between, on the one hand, conservationists and indigenous rights activists, who regard the indigenous islanders as essential guardians of their reserve, located in one of the few remaining regions of biodiverse rainforest, and, on the other hand, local politicians, who voice the demands of various settler groups desiring to get access to precious forest resources and land for the sake of 'development'.

The second dialectics this contribution is going to analyse concerns forest policies and resource use, too, but concentrates on forest labour and questions of place-making in ecological and social margins. Since the Andaman Islands had been colonized permanently from 1858 onwards, spatial expansions of settlements and infrastructure went hand in hand with forest clearance and a rapidly expanding timber export industry. Thus, capitalist development in the Andamans relied heavily on the exploitation of forest resources. To meet the existing demand for suitable forest labour, the British transported labour migrants from the Central Indian Chotanagpur plateau to the islands from 1918 onwards; these labourers, called Ranchis, belonged to a large array of different tribal groups. These tribal communities had lived an anarchic way of life – which may be characterized, among other qualities, by a political system of self-governance and a self-sufficient mode of production – in their central Indian homelands until being displaced by colonial forest and grazing policies in the nineteenth and twentieth century (cf. Sundar 2007).

After independence from Great Britain in 1947, the Indian administration continued to contract Ranchis to clear the tropical rainforests, both for the once thriving timber industry and to develop the infrastructure for the settlement of diverse migrant communities from the subcontinent. As they were not given any land to settle down, many Ranchis encroached plots of cleared forest and reconstructed an anarchic lifeworld based on the principle of using locally available forest resources and on the primacy of evading the state and keeping it at a distance. The illegality of their encroachments was largely overlooked by the state until the turn of the millennium, when Indian forest policies underwent a major paradigm shift from exploitation to conservation. Since the Supreme Court ordered the eviction of all encroachers of forest land in May 2002 – according to the rationale that they are harmful to the forest ecology – the Ranchis' villages are under constant threat of eviction. While evictions from their plots have, to a large extent, not been carried out, this policy shift has brought with it crucial implications for the Ranchis counter-hegemonic strategies aiming to 'conserve' their own 'anarchy'. An anarchist political ecology debating the impact of such strategies may gain from considering the emic understanding of social and cultural anthropology, a discipline which has traditionally dealt with societies living in anarchy but often refused to name their anarchy correspondingly. Seeking to lay out a theoretical groundwork for the ensuing ethnographic discussion in this chapter, I am going to enter the discussion by elaborating on how the anarchic trajectory of humanity intersects both with the political movement of anarchism and with anthropology as an academic discipline.

ANARCHY, ANARCHISM AND ANTHROPOLOGY

The *long durée* of humanity's political history could be narrated as the history of the absence of state domination. Humans have governed themselves for the longest part of their existence on planet Earth (Graeber 2008, 65; Scott 2009, 3). In contrast, state rule may be regarded as a comparatively recent invention, which, for some thousands of years, affected mere fragments of the world's population. However, along with the birth of the modern state in the late eighteenth and early nineteenth century, the state idea and its institutions have spread across the globe and have, ever since, continued to penetrate and subdue previously self-governed communities. As a result, the state idea came to assume an unquestioned and seemingly unquestionable discursive status as producer of order, equality and justice, deemed necessary in order to avoid and prevent the supposedly always impending threat of chaos, injustice and violence, which are, according to the hegemonic state narrative, the features of a state of statelessness – a condition that has once and again been wrongly depicted with the term 'anarchy' (Graeber 2008, 10). Contradicting such a dubious representation, a short glance at history books will prove that inequality, marginalization, domination, dispossession, ethnocide and genocide cannot, in most cases, be attributed to the absence of the state; they must rather be regarded as an effect of intensification, expansion, disintegration or transformation processes of state rule itself.

Since ancient times, the term 'anarchy' (*an-archía*: without rule) describes an idealized condition of freedom from rule. Anarchy, however, does not imply the complete absence of hierarchy or conflict, but the absence of institutional structures or rule and domination designed to avoid and prevent any outsider or member of a community or society to exert dominance without their consent. An anarchic society can therefore be conceptualized as an ordered society without government (or similar dominating or exploitative structures), in which order is based on the maximum self-reliance of all its members. Anarchic communities regard all their members as mature and thus demand their participation in basic democratic decision-making processes as well as in the implementation of these decisions seeking to benefit the whole society.

These core principles inspired political theories of anarchism, which evolved in the last decades of the eighteenth century, and formed a social movement in the course of the nineteenth century (Morris 2014, 63–64). The anarchist movement mobilized its followers largely in already industrialized or industrializing societies with the explicit goal to organize against or to strategically circumvent state rule. The political ideology of anarchism (cf. Proudhon 2011; Chomsky 2013) has, however, so far not exerted much influence on those anarchic societies inhabiting the geographical and social peripheries of the Global South: herding nomads, foragers and subsistence-oriented peasant societies that are often described as tribal, aboriginal, indigenous, autochthon, native, first nations and so on; these are, in most cases, affected by political, socio-economic and cultural marginalization within state societies.

For the most part of its history as a discipline, anthropology has extensively studied such 'non-state' peoples' mode of production, cultural norms, values and practices, kinship and political organization, language, ritual, religion and legal institutions in order to come to generalizing conclusions on the human condition.[3] The core values that inspired the forms of political institutions encountered by anthropologists among non-state (egalitarian, segmentary, akephal, polykephal, et cetera) societies show a great overlap with anarchist principles such as individualism, equality, decentralization of rule, cooperation, mutual help, basic democratic forms of decision-making as well as the adherence to strategies of fighting or avoiding domination from outside the community. The majority of groups living an anarchic way of life does, however, neither know about the political ideology of anarchism nor identify with it (cf. Graeber 2008; Morris 2013).

Acknowledging a thematic and empirical conjuncture between various forms of anarchy studied by anthropologists and theories about anarchy developed by anarchists, I propose to take a closer look at the ways in which anthropological studies of anarchic societies and the political ideology of anarchism have taken influence on each other. Anarchists like the geographer Peter Kropotkin (2009) extensively referred to anthropological studies in their work and inspired his friend and follower, the British anthropologist A. R. Radcliffe-Brown who was called 'Anarchy Brown' before he became the prominent founder of British structural-functionalism (Goody

1991, 1; Perry 1978, 61). Some anthropologists referred to anarchist vocabulary when describing non-state societies: for example, Edward Evans-Pritchard (1940, 6) termed the political system of the Nuer in Sudan, which lacked of institutionalized authority, as 'ordered anarchy'; Christian Sigrist (2005) described the absence of formal rule among segmentary lineage societies in Africa as 'regulated anarchy'; in his theory of peasant economies, Marshall Sahlins (1974) called the domestic mode of production, in which production and consumption are organized at the household level, a 'species of anarchy'; especially Pierre Clastres' 'Society Against the State' (1987) relies on anarchist ideas when depicting indigenous Amazonians' strategies of averting unwanted, centralized political power from within and outside. In general, one can assess a growing impact of anarchist theory on anthropological, sociological, historical and geographical research in the last three decades such as in the work of Amborn (2016), Barclay (2009), Gibson and Sillander (2011), Graeber (2008), Ferretti et al. (2017), Macdonald (2008, 2010, 2011), Morris (1986, 2013, 2014), Scott (1985, 2009), Springer (2016) and White et al. (2016).

Research on anarchic societies may be characterized by a broad interest in the political, cultural, socio-economic and ecological strategies striving to preserve or retrieve partial or complete autonomy from the state. These practices assume primary importance in the light of the growing expansion of states and markets in previously self-governed and self-sustaining spaces inhabited by anarchic communities. According to David Graeber (2008, 54), however, expanding state institutionalization does not necessarily imply an undoing or vanishing of anarchic spaces, but rather a coexistence of different forms of statehood and anarchic modes of self-governance. Consequently, anarchy must be regarded as an ontological condition of humanity implying that humans adopt creatively to different forms of domination, and that they always find strategies to circumvent or sabotage these (Zehmisch 2017a). Anarchist political ecology has the potential to carve out the contours of such forms of resistance by focusing on various ecological practices serving to maintain partial or complete political autonomy from the state.

THE STRUGGLE BETWEEN ARCHY AND ANARCHY IN THE ANDAMANS

The historic process of state institutionalization in the Andaman Islands in the last 150 years can be read as a constant struggle over the control of territories and resources. Lastly, it may also be conceptualized as a struggle between *archy* (rule) and anarchy. When the British Empire installed a penal colony around today's capital Port Blair in 1858, a process of gradual erasure of forest and marine resources, the economic and ecological base of the 'Andaman Islanders',[4] unfolded as a byproduct of colonization. Transferring convicts from overcrowded jails on the Indian subcontinent to the islands in order to punish and reform them served to establish a permanent outpost in the Indian Ocean by utilizing forced labour (Sen 2000; Vaidik 2010). The convicts' task was to clear primary forests, to drain swamps and to develop the infrastructure necessary to accommodate more than 10,000 persons (Mathur 1985, 37). Another central element of spatial transformation was to convert the 'wild' jungle into a settled agricultural zone by rehabilitating convicts as self-supporting farmers on cleared forest lands. The establishing of a full-fledged colony in the primary forests of the Andamans may thus also be interpreted as a settler-colonial project aiming to subdue and transform 'nature' and 'savages' and to eventually replace and eliminate the indigenous population (cf. Wolfe 1999, 163; Sen 2016, 75). Such an 'ecological warfare' (Sen 2010, 65) against the jungle and its human and more-than-human inhabitants implied minimizing the autonomy and autarky of the Andamanese and had genocidal and ethnocidal consequences. Constructing the indigenous islanders as 'savage others' – an anarchic counterpart to the colony living across the sociopolitical, cultural and ecological frontier of 'civilization' – to the state served to stabilize the subject position of the colonizers as racially and culturally superior, too.

Of course, the Andamanese tried to defend themselves against these intrusions into their territory and against such violent appropriations of their means of sustenance and survival. A coordinated full-scale attack by the Aka-Bea-Da, a Great Andamanese tribe, in 1859 ended in a massacre of around 400 warriors. The British troops had been prepared for the attack because of a warning given by a runaway convict, who had lived among the Great Andamanese for more than a year before betraying them (Pandya 2009, 83). A memorial in the centre of Port Blair town remembers this event nowadays as 'the Battle of Aberdeen'. After their 'pacification', some Great Andamanese worked as trackers of runaway convicts and participated in punitive expeditions against the hostile Jarawa (Pandya 2010, 21).[5] Long-standing violent conflict and competition for resources caused the retreat of the remaining Great Andamanese, Jarawa and Onge into more peripheral forest areas in order to avoid interaction with the colonizers that threatened their autonomy and sustenance. The British, however, did not need to conquer the whole Andamanese territory but rather confined themselves to their 'civilizing mission' and medical interventions (Sen 2010, 144). These entailed the isolation of Andamanese children in so-called 'Andaman Homes' in order to missionize and educate them. Resulting from such close contact with colonizers and convicts in the penal settlement, diseases like measles and syphilis spread into the forests and extinguished the vast majority of nine Great Andamanese sub-tribes (Chandi 2010a, 17).

The policy of stretching the frontier into indigenous lands was combined with the expansion of commercial forestry too. Since the last decades of the nineteenth century, timber was the major export commodity of the islands (Lal 1976, 33–34; Devaraj 2001, 157). The famous Andaman Padauk (*Pterocaopus dalbergioides*) was shipped to Europe and North America (Sen 2000, 19); some of it supposedly adorns even the interior of the Buckingham Palace in London. Large quantities were also transported to the Indian mainland for the construction of railway sleepers (Zubair 2011). The task of logging the abundant and biodiverse forests had been first handed over to convicts. After some decades, however, convicts came to be regarded as supposedly not fit enough for arduous forest labour. That's why from

1918 onwards, the administration started to transport specialized forest labourers to the islands, the Ranchis. These aboriginal migrants from the Chotanagpur region in Central India had gained their name from being contracted as 'hill coolies' by the Catholic Labour Bureau in the city of Ranchi. The Ranchis belonged to different tribes: the majority were Oraon, Munda and Kharia; subsuming various aboriginal groups under the label 'Ranchis' implied creating a subaltern class of aboriginal migrant labourers in an overseas location. The British had classified aboriginal labourers from Chotanagpur as 'first-class coolies', because they were assumed to be 'docile', 'hard-working' and 'racially fit' to endure adverse climatic and ecological conditions (Ghosh 1999, 29–32). It was expected that the Ranchis would adjust well to the harsh Andaman forest environment (Raju 2010). Beyond that, it can be assumed that these nominal aboriginal peasants who hunted and gathered were intentionally given the task to clear forests that were still inhabited by the Andamanese. Being considered less 'savage' than the Andaman Islanders but considerably more 'primitive' than convict settlers from non-tribal communities, the Ranchis were an integral part of the ecological warfare against the anarchy of the jungle. Their labour therefore enabled a semantic and ecological transformation of 'unproductive' and 'wild' forests into 'productive' settler colony spaces (cf. Zehmisch 2016b).

After independence from Great Britain in 1947, the Indian administration started to shift various subaltern groups from all over South Asia to the islands, which contributed to the shaping of a multi-ethnic island society, called 'Mini-India', on former indigenous lands (Zehmisch 2017a). Hindu refugees from East Bengal, Tamil and Telugu Repatriates from Burma and Sri Lanka, and landless people from South and Central India were settled by the government, while huge number of migrants from all over the subcontinent came on their own without government support. Such population movement also caused the massive encroachment of 'free spaces' of forest land – that were in most parts vacated by the Andamanese and thus conceptualized as *terra nullius*. As a result of these migration movements, the Andaman population has grown to currently more than 400,000 people.

The expansion of the frontier into forests for settlements, road construction and the growing timber industry after independence caused the remaining indigenous islanders to engage with the state, to retreat further in the forest or to violently defend their territory. Here, it makes sense to be reminded of Pierre Clastres' argument that war in non-state societies functions to uphold a people's 'undivided totality', and thus their autonomy, collective sense of community, territory and enmity against others (2008, 67). Presently, approximately 750 indigenous islanders, divided in four ethnic groups with significant cultural and linguistic differences, continue to live in the Andamans. Numerical decline among these four groups was highest among those who had been pacified earlier: the Great Andamanese, who comprise today around fifty persons, and the Onge (around 100 persons living on Little Andaman) have largely given up their hunting and gathering mode of production, and survive on the basis of rations provided by the administration. In contrast, the fifty to hundred Sentinelese, who attack anyone entering their remote island, have so far succeeded in maintaining their autonomy and autarky.

The Jarawa (around 450 persons) are currently experiencing drastic sociocultural and economic transformations since they have given up their violent resistance against the colonization of their territory in 1998 (Sekhsaria/Pandya 2010, 8). The earlier violence exerted against encroaching, poaching and intruding outsiders had served to protect the ecologically unique forest-scape of the reserve and thus the resources and autonomy and autarky of the Jarawa. Since incidents of armed conflict and reprisal have gone down, and as groups of Jarawa came to visit their neighbours and made peace with them (Chandi 2010a, 15), settlers, migrants and government servants have increasingly entered their tribal reserve (declared in 1956) in order to poach, fish, log hardwood or to extract precious resources from the forest or the sea shore. In addition to that, a barter system between the Jarawa and some settlers has developed, in which alcohol, food and other commodities are exchanged against game and honey. Extending their foraging activities, the Jarawa now frequent villages around the reserve in order to collect crops and fruits from the gardens and planta-tions of settlers (Chandi 2010b, 15). Based on the assumption that they don't find enough means of subsistence in the forest due to the pressure on forest resources, the administration has started to provide them rations (rice, flour, sugar, salt, lentils). This has brought about a significant change to their previously protein-rich diet (cf. Zehmisch 2017b).

Official policies towards the Jarawa have undergone drastic change too: since the 1970s, their territory had been colonized by constructing the Andaman Trunk Road (ATR) directly through the reserve and by instituting twenty-six check posts of the so-called 'bush police'. This policy had the broad aim to contain the violent resistance of the Jarawa and to gain control over the reserve. In stark contrast, there is currently a strong demand and will to enforce the Jarawas' isolation from outsid-ers with strict rules and regulations. While they had earlier been perceived as part of 'nature' that needed to be colonized and transformed, they are nowadays regarded as vulnerable and in need of state patronage because of their perceived closeness to 'nature'.

This paradigm shift towards augmenting the Jarawas' protection may have been influenced by national and transnational activism for indigenous rights (cf. Ghosh 2006). These activists place the Jarawa firmly within the transnational discourse on indigeneity – a discourse to which the Jarawa themselves don't subscribe to (Pan-dya/Mazumdar 2012, 51). Representing the indigenous islanders along with their 'pristine' ecological environment as a vulnerable part of 'nature', which is threatened by civilization, modernity, development and migration, they tend to evoke a rather contemporary version of the earlier colonial trope of the 'noble savage', the 'ecologi-cally noble savage' (Hames 2007). Here, activism for indigenous rights goes hand in hand with the conservation paradigm. The Jarawas' cause thus links the onslaught of 'civilization' on vulnerable indigenous peoples with depletions of a unique eco-system such as that of the Andamans. A study published by UNESCO expresses the discourse and conflict over the territory in the following way:

The Jarawa Reserve, being the last repository of the variety of ecosystems specific to the Andaman Islands, is increasingly under scrutiny from by conservationists and natural resource users/ exploiters. This storehouse of biodiversity and its surrounding waters allows for exploitation to be commercially viable as many regions close to settlements have now been extensively degraded. (Chandi/ Andrews 2010, 49)

Further, it has been acknowledged that 'the patchwork of ecological niches within this territory are the last pure examples of the biological diversity of the Andaman islands' (Chandi 2010b, 35). This is also corroborated by the fact that 'the last regions that contain the "giant Andaman evergreen forest" are now restricted to the Jarawa Reserve' (Singh 2003, in: Chandi/Andrews 2010, 50). These political ecology arguments are, moreover, part of a larger civil society debate between a set of local politicians, deeply mired in old-fashioned perceptions of 'modernity' and 'progress', and activists promoting conservation of the unique Andaman environment along with its indigenous inhabitants. The politicians claim to speak for local entrepreneurs and concerned settlers living in vicinity to the Jarawa reserve, who perceive the Jarawa as a hindrance for the region's development (Chandi 2010a, 15). Their major demand is to end the policies of isolating the Jarawa, assimilate, modernize and educate them, and to exploit the valuable resources of their reserve. The opposite position is taken by both international and national activists who articulate voice for the Jarawa in the global public sphere and who demand from the authorities to more effectively protect the indigenous population along with the vulnerable forest and marine environment (Zehmisch 2017a). The most prominent point of debate, which had once and again been emphasized by the activists, is the Andaman authorities' contempt of a Supreme Court order from 2002, which had passed a judgement that those parts of the ATR crossing the Jarawa reserve must be closed (cf. Sekhsaria 2007, 84–86). The road has, however, so far not been closed, and tourist and local traffic continue to ply through the reserve; in addition to that, plans for the construction of a railway line through the Jarawa reserve, connecting the capital of Port Blair with North Andaman, have been issued. As these plans of yet another large infrastructure project signify utter neglect of vulnerable ecosystem of the islands and its indigenous population, it will surely elicit further organized resistance by environmentalists and indigenous rights activists.

While such drastic state interventions have continuously minimized the anarchy of the indigenous hunting and gathering communities since almost one and a half centuries, one can also observe that certain state interventions paradoxically serve to retain a certain autonomy and autarky from the state itself. For example, special legislation allows the indigenous islanders to hunt and forage and to freely use all forest resources; further, they are not liable to be punished by the law for crimes like murder and theft. Thus, the indigenous islanders are on the one hand kept in a cage of primitivism – relegating them to a 'sub-human' status by not applying the concept and duties of citizenship on them (cf. Sen 2016) – which, on the other hand, may enable them to at least partly maintain anarchic way of life and mode of production.

THE PARADOX OF CONSERVATION

The anarchic history of the Ranchis of Andaman can be told as an alternative, sub-altern history of island colonization. Transported to islands because of their special abilities as forest labour, they were the major actors of colonization – building roads, dams, jetties and airports, and providing logs to the timber industry – but their contri-butions to the 'development' of the Andamans have never been recognized by the state (cf. Zehmisch 2016b). While Ranchis have kept coming to the islands since 1918, the major bunch migrated after independence, when the Indian government continued the recruitment of these precarious *Adivasis* (indigenous people of the Indian subcon-tinent, literally: first dwellers) in collaboration with the Catholic Labour in Ranchi and private contractors. Their recruiters and the authorities did not imagine them as potential settlers but treated them as precarious and footloose labour power. Contrary to official expectations that they would return to their homelands after their contracts had ended, many Ranchis encroached remote plots of forest land, which they had once cleared for the purpose of timber extraction and stayed on (cf. Zehmisch 2016b).

Most former labourers returned to a peasant mode of production that was cru-cially influenced by the legacy of autonomy and autarky in their homelands on the subcontinent, where no concept of private land ownership had existed before it was colonized (Bates/Carter 1992). Encroaching available land for housing, gardening or paddy cultivation was a common practice applied in a diasporic context. As these subaltern *Adivasis* had anyways no other means of income after their contracts had terminated, they returned to a subsistence-oriented mode of production which they – and generations of their ancestors – had practised in their homelands: small-scale agriculture and gardening, livestock herding, as well as hunting and foraging in the surrounding forests. As there has always been a constant need of cash for consump-tion of basic amenities such as clothes, alcohol, cigarettes, household items and food rations, too, young family members have also been working as wage labourers in larger villages and towns in order to reimburse money to their relatives living on encroached plots.

While many of the currently around 50,000 Ranchis continue to live in the second or third generation – since thirty to fifty years – in encroachments, few of their plots have been regularized and thus legalized; this stands in stark contrast to some members of other ethnic communities, who have managed to legalize their encroached plots due to political and bureaucratic patronage and/or enough capital to pay sufficient bribes to bureaucrats (cf. Zehmisch 2016a). Encroached villages have usually no streets, electricity, canalization and few properly functioning schools. A combination of these factors has inhibited the majority of Ranchis to escape the shackles of poverty and marginalization. These conditions, in turn, may have moti-vated them to reconstruct a quiet, self-governed life in economic sustenance far away from the oppressing leverage of the dominant state.

Around the turn of the millennium, however, the Ranchi encroachers' legal and political situation came to be crucially influenced by a remarkable shift in envi-ronmental policies proposing the conservation of the very same resource, which had earlier caused the authorities to bring Ranchi labourers to the islands: forests. Characteristic of the slowly growing awareness of anthropogenic impacts on Planet Earth's climate among activist and progressive government circles in India, the above-mentioned order of the Supreme Court of India in May 2002, which had demanded a closure of the ATR through the Jarawa reserve, also came to signify a major paradigm shift in forestry from resource exploitation towards environmental conservation (Sekhsaria 2007, 41). The Supreme Court order was based on a report submitted by the commissioner Shekhar Singh who had conducted a survey about the condition of the island ecology and the protection of the Indigenous islanders (ibid.). His recommendations included, among others, a ban on logging naturally grown trees, a ban on timber export and, most crucial for the local population, the removal of illegal encroachers (ibid.). The contents of the recommendation caused long-lasting and heated public debates, which is well expressed in the following statement:

> The regulations imposed by the Supreme Court have created a perception of injustice towards the settlers, further polarizing the settlers against those pressing for environmental concerns, which are intimately linked to the rights of the indigenous groups in the islands. (Venkateswar 2007, 4–5)

Though only partially implemented, the order had tremendous impacts on official policies seeking to balance between crucial livelihood issues of the majority of the population and legitimate demands to protect the vulnerable Andamanese and the biodiverse tropical environment. For example, the prohibition of timber export made workers redundant and caused political tensions, but it also contributed to considerable regenerations of secondary forests.

The situation of the Ranchi encroachers after the Supreme Court order deteriorated their relationship to the state, and confronted them with severe social and legal problems. The same order which had demanded the closure of the ATR through the Jarawa reserve in the name of conservation and protection, now endangered the livelihood and housing of the aboriginal migrants from Ranchi, who had settled in those biodiverse forestscapes they had earlier been ordered to log for timber exports. According to the order, all encroachers should have been removed from forest land within a few months. However, against the expressed content of the order, large-scale evictions of encroachers have not taken place so far because of the humanitarian tragedy it would have entailed, as a Forest Department official once told me. There had been, however, few selective and arbitrary evictions, and a large number of gardens and plantations of encroachers were cut down and destroyed with the help of working elephants. The intention behind such destruction of encroachers' means of subsistence was probably the calculation that they would leave the encroachments. Resisting these attacks on their livelihood, most Ranchis I had met during several spells of fieldwork in 2011 and 2012 went on to rebuild their gardens and plantations, and struggled for additional income through precarious wage labour to compensate for the loss of crops (cf. Zehmisch 2017a). When I confronted my Ranchi interlocutors with the conservation agenda due to which they should be removed from their new homelands, they replied that they were actually conserving the environment, as they had planted fruit trees and gardens on earlier cleared plots that had, at the time of their settlement, been largely barren of trees.

The legal limbo, in which Ranchi encroachers currently live in, continues to criminalize their subsistence-oriented mode of production and threatens their political autonomy from the state. Looking at the Ranchis from a perspective that acknowledges their agency, however, it appears safe to claim that the insecure status of their occupied lands also indirectly enables them to live in greater anarchy because the state and the market do hardly interfere in marginal encroachment villages through governing, welfare, commercial or entrepreneurial activities. For example, schools are the only existent government institutions in encroachments, though they do rarely function as teachers are usually absent most of the time. Thus, there are usually, apart from forest department outposts in few villages, hardly any government servants present, who could try to impose their authority over Ranchi villagers. Interlocutors living in encroachments repeatedly stated to be glad that state officials did rarely come to their remote villages, because they did hardly work for the community but usually try to violently extract bribes and resources or to harass Ranchi women.

To return to the larger argument about the anarchic life world of the Ranchis, I would like to give some examples in which ways their return to a mode of subsistence in marginal forest spaces may be interpreted as a conscious evasion from state influence (cf. Scott 2009). Their active avoidance of the state and the majority society of the Andamans is confirmed by consensual and basic democratic decision-making processes regarding all matters that directly concern the village community. One prominent example, narrated to me by members of an encroachment village, was their circumvention of the electoral mechanism of the state in communal elections, by internally deciding on a village representative in a basic democratic process – only to cast the vote for the same person in the upcoming elections.

The Ranchis' development of anarchic means of self-governance went along with a remarkable adaptation to the forest environment. In broad generalization, the Ranchis have 'imported' their cosmological beliefs in an 'animated' environment that encompasses both human, more-than-human and spirit actors, from their homelands in Central India. Beyond relating to ecological surroundings, my interlocutors regarded the spirits inhabiting the forest and marine environment as integral constituents of their lifeworld. Their extensive knowledge of the Andaman landscape, flora and fauna may be regarded as a result of their emplacement in new ecological spaces. When these labourers once logged down the biodiverse primary forests, they supplied themselves partly through gathering and hunting. Coming largely from forest-dwelling backgrounds, they had a strong sense of the environment due to which they gradually discovered the qualities of island species – probably through a trial-and-error system. As a result, within few generations after their migration to the Andamans, many Ranchis I met were able to identify a large variety of local plants and animals and knew how to utilize them as both nutrition and indigenous medicine. This knowledge may have been transmitted across generations. In remote forest areas, the Ranchis' non-professional 'jungle doctors' (*vaid*) are often the only means of medical care; they show, however, considerable success when treating a large range of ailments such as snake and centipede bites, inflammation, skin diseases and spirit possessions with wild-growing forest plants in elaborate rituals.

Moreover, their joint settlement of encroached forest spaces led to close neighbourly relations among the village community, which may have evolved on the basis of similar conditions of survival through subsistence in an equally novel environment. As a result, Ranchi village communities are usually governed by ethics of sharing and common solidarity that are opposed to the principle of market economy (cf. Macdonald 2011). For example, all members of the village community provide mutual aid to each other when it comes to the planting and harvesting of crops, the construction of houses, stable, gardens and fences. These forms of communal, reciprocal labour are usually called *madad* (help). All participants are remunerated with a nominal sum but provided with free *handia* (rice beer) and a feast by the supported family. As the village community confirms and celebrates its unity in transition rituals such as betrothals, marriages, baptisms and funerals, the preparation of feasts is usually supported by members of the community who hunt, gather, slaughter, cook, construct shelters and prepare *handia*.

Corresponding to Sahlins' (1974) famous descriptions of hunter-gatherers as 'original affluent societies', who find 'material plenty' in their environment, Ranchi interlocutors also expressed with certain pride and relief to inhabit geographically marginal zones where the forest and marine environment provides them with resources for their sustenance: 'sweet' drinking water; non-polluted air to breathe 'freely'; game and a large variety of wild-growing fruits, ferns and vegetables; building material for houses, fences and sheds; crops from gardens, plantations and paddy fields; fodder for livestock such as cows, buffaloes, pigs, chickens and goats; and fish and seafood from the shores and creeks. It is thus no exaggeration to state that most Ranchis living in remote forest encroachments have access to sufficient resources that enable them partial autarchy and autonomy from external sources of income as well as the influence of bureaucrats.

While the Ranchis' anarchy is currently under threat by government interventions seeking to enhance the conservation of the unique Andaman environment, the lack of implementation of the Shekhar Singh Commission's recommendations provides them with a strongly desired absences of the state and considerable space to exert their independence, self-governance, mutual aid and sustenance. This precarious equilibrium is, however, psychologically devastating for the encroachers themselves, as their everyday life is shaped by a constant, theoretical threat of eviction. Further, in case the government adheres to its core disciplinary technology of rule, the enforcement of its sovereignty, as it continues to do in many parts of the subcontinent, and decides to prioritize the preservation of forest resources over the livelihood of its human inhabitants, then this anarchic *status quo* is liable to change drastically. Environmental and indigenous rights activists should therefore be attentive to find a balance between their agenda and that of the affected communities – a balance that considers both the demands of restoring the diversity of the forest, and, at the same time, providing the forest-dwellers with means to conserve their anarchic way of life as an integral part of the forest. One way to find such a balance may be listening to the Ranchi's articulations about their needs and imaginations of a 'good life' and to closely observe their practices of conserving both the environment and their anarchy.

CONCLUSION

The conservation of anarchy has come to assume multiple meanings in the context of the Andaman Islands. It can, albeit in different ways, be applied to both examples discussed in this chapter. Contemporary state policies of protection and welfare are paradoxically instrumental in partially maintaining the very anarchy of the Andaman Islanders that the state had earlier sought to minimize through violent and forceful measures.

In the case of the Ranchis, this argument can only be made in a more indirect way. The Ranchis' acts of establishing a livelihood and existence on encroachments of forest or revenue land outside the settled zones may not only be read as threats to the unique Andaman environment. Their sustained forms of self-rule, external to the hegemony of state governance, can, moreover, be interpreted as implicit or explicit acts of resistance to attempts of the state to control its margins. Thus, state policies of environmental conservation may be regarded as an indirect, legal attack on the Ranchis' sustenance. However, as the maintenance of the *status quo* has not significantly altered their subaltern positionality as marginalized encroachers of forest land – who could be evicted, but have not been so far due to reasons of governance – the state also indirectly maintains the status of the Ranchis as living in ecological and socio-economic margins, where they can continue a partially self-governed and self-sufficient way of life based on those forest resources, which the Supreme Court order aims to protect from the impact of encroachers.

For state populations that have embraced hegemonic ideas like the rhetoric of 'civilization', such anarchist autonomy, as described in this chapter, remains an abstract and distant utopia. This applies certainly to many indigenous rights and environmental activists, too. While they fight for global sustainability and social justice, many conceive anarchic autonomy as evolutionary condition that existed either in the 'deep' past or as a condition of some 'not yet civilized relics' of this past. Such romanticizing or 'primitivist' positions lead activists to construct indigenous peoples and their hitherto unexploited territories and markets as in need of 'salvaging' and protection from the brutal and ruthless onslaught of capitalist modernity. Such stances might explain why the struggles of indigenous peoples, if politicized, are liable to create strong emotional support among certain sections from the global North and civil society members from the South. Exoticizing these others in need of 'benevolent' support by civil society members, many activists do, however, not adequately recognize the agency of anarchic communities. As a consequence, supporters of their causes often fail to understand and thus to advocate for the conditions of anarchy that many marginalized peoples on this planet wish to maintain.

NOTES

1. In this chapter, the term 'indigenous' will be utilized as a contextual, relational and polit-ically loaded term, which is consciously used by indigenous and non-indigenous actors alike in order to claim protection, rights and recognition from the state. Indigeneity thus may be understood as meandering between essentialist ascriptions of identity by non-indigenous actors and processes of appropriation, reinterpretation, rearrangement and camouflage of categories of indigenous-ness by those who are signified as indigenous themselves (Zehmisch 2017a, 287).

2. Together with the Nicobar Islands, the Andamans constitute the Indian Union Terri-tory Andaman and Nicobar Islands in the Bay of Bengal, comprising all together 572 islands, reefs and rocks (Sekhsaria 2009, 256). The islands are located in geographical vicinity to Southeast Asia, at a distance of more than 1,000 kilometres from the Indian subcontinent. Due to their strategic importance, this former British colony is now under direct rule of the Indian central government in New Delhi.

3. See Maine (1861), Morgan (1877), Radcliffe-Brown (1922), Fortes and Evans-Pritchard (1940), Leach (1954) and Sahlins (1974).

4. Among anthropologists, the term evokes association with Radcliffe-Brown's structural-functionalist classic 'The Andaman Islanders' (1922). In this chapter, I am using the terms 'Andaman Islanders' and 'Andamanese' interchangeably to refer to all ethnic groups of indigenous islanders. The colonial administration had estimated the number of all indigenous islanders to be around 6,500 (Pandya 2009, 74). The nine tribal subdivisions of the Great Andamanese counted around 5,000 people, the Jarawa around 600, the Onge 700 to 1,000 and the Sentinelese 50 to 100 people.

5. Jarawa means 'stranger' in Aka-Bea-Da, one of the nine Great Andamanese dialects. It indicates a longstanding, probably antagonistic relationship between both groups (Chandi 2010b, 32). The Jarawa refer to themselves as Ang (people/humans). The languages of the Jarawa and Onge belong to a separate language group from the Great Andamanese and differ in grammar, vocabulary and sound system (Abbi 2006).

REFERENCES

Abbi, A. (2006). *Endangered Languages of the Andaman Islands*. München: Lincom Europa.
Amborn, H. (2016). *Das Recht als Hort der Anarchie. Gesellschaften ohne Herrschaft und Staat*. Berlin: Mattes & Seitz.
Barclay, H. (2009). *People Without Government: An Anthropology of Anarchy*. London: Kahn & Averill.
Bates, C., & Carter, M. (1992). Tribal Migration in India and Beyond. In *The World of the Rural Labourer in Colonial India*, edited by Gyan Prakash. Delhi: Oxford University Press: 205–247.
Chandi, M. (2010a). Colonisation and Conflict Resolution: Learning from Reconstruction of Conflict Between Indigenous and Non-Indigenous Islanders. In *The Jarawa Tribal Reserve Dossier – Cultural and Biological Diversities in the Andaman Islands*, edited by Pankaj Sekh-saria & Vishvajit Pandya. Paris: UNESCO: 12–17.
———. (2010b). Territory and Landscape around the Jarawa Reserve. In *The Jarawa Tribal Reserve Dossier – Cultural and Biological Diversities in the Andaman Islands*, edited by Pankaj Sekhsaria & Vishvajit Pandya. Paris: UNESCO: 30–42.

Chandi, M., & Andrews, H. (2010). The Jarawa Tribal Reserve: The 'Last' Andaman Forest. In *The Jarawa Tribal Reserve Dossier – Cultural and Biological Diversities in the Andaman Islands*, edited by Pankaj Sekhsaria & Vishvajit Pandya. Paris: UNESCO: 43–53.

Devaraj, P. (2001). *Forests of Andaman Islands*. New Delhi: IBD.

Evans-Pritchard, E. E. (1940). *The Nuer – A description of the Modes of Livelihood and Political Institutions of a Nilotic People*. Oxford: Clarendon Press.

Ferretti, F., Toro, F., Barrera, G., & Ince, A. (Eds.). (2017). *Historical Geographies of Anarchism – Early Critical Geographers and Present-Day Scientific Challenges*. Milton Park, Abingdon, Oxon: Routledge

Fortes, M., & Evans-Pritchard, E. E. (Eds.). (1940). *African Political Systems*. Oxford: Oxford University Press.

Ghosh, K. (1999). A Market for Aboriginality – Primitivism and Race Classification in the Indentured Labour Market of Colonial India. In *Subaltern Studies X – Writings on South Asian History and Society*, edited by Gautam Bhadra, Gyan Prakash, & Suzie Tharu. Delhi: Oxford University Press: 8–48.

———. (2006). Between Global Flows and Local Dams – Indigenousness, Locality, and the Transnational Sphere in Jharkhand, India. In *Cultural Anthropology*, Vol. 21, 4. Berkeley: University of California: 501–534.

Gibson, T., & Sillander, K. (Eds.). (2011). *Anarchic Solidarity – Autonomy, Equality, and Fellowship in Southeast Asia*. New Haven, CT: Yale University Southeast Asia Studies.

Goody, J. R. (1999). Anarchy Brown. *Cambridge Anthropology* 21, no. 3: 1–8.

Hames, R. (2007). The Ecologically Noble Savage Debate. *Annual Review of Anthropology* 36: 177–190.

Heidemann, F., & Zehmisch, P. (Eds.). (2016). *Manifestations of History – Time, Space, and Community in the Andaman Islands*. Delhi: Primus.

Kropotkin, P. (2009/1902). *Mutual Aid – A Factor of Evolution*. London: Freedom Press.

Lal, P. (1976). *Andaman and Nicobar Islands – A Regional Geography*. Calcutta: Anthropological Survey of India.

Latour, B. (1993). *We Have Never Been Modern*. Trans. Catherine Porter. New York: Harvester Wheatsheaf.

Leach, E. R. (1954). *Political Systems of Highland Burma – A Study of Kachin Social Structure*. London: Bell.

Macdonald, C. (2008). *Cooperation, Sharing and Reciprocity (Sharing without Giving, Receiving without Owing)*. https://docs.google.com/viewer?a=v&pid=sites&srcid=ZGVmYXVsd GRvbWFpbnxjaGFybGVzamhtYWNkb25hbGRzc2l0ZXxneeD phNDcyZTcyOTMwY WM2MzM.

———. (2010). *Can Anarchism be a Critical Point in the New Anthropological Imagination? Contributions to Anarchy and Anarchism to Social Theory*. https://docs.google.com/viewer?a=v&pid=sites&srcid=ZGVmYXVsdGRvbWFpbnxjaGFybGVzamhtYWNkb25hbGRz c2l0ZXxneeDo2ODM4YTY3NGYzNGE4ZGNm [08.02.2018].

———. (2011). A Theoretical Overview of Anarchic Solidarity. In *Anarchic Solidarity: Autonomy, Equality, and Fellowship in Southeast Asia*, edited by Thomas Gibson & Kenneth Sillander. New Haven: Yale Southeast Asia Studies: 17–39.

Maine, H. S. (1861). *Ancient Law – Its Connection with the Early History of Society, and Its Relation to Modern Ideas*. London: John Murray.

Mathur, L. P. (1985). *Kala Pani – History of Andaman & Nicobar Islands with a Study of India's Freedom Struggle*. New Delhi: Eastern Book Corporation.

Morgan, L. H. (1877). *Ancient Society*. New York: Henry Holt & Co.

Morris, B. (1986). *Forest Traders – A Socio-Economic Study of the Hill Pandaram*. London: Athlone Press.

———. (2013). *Anarchism, Individualism and South Indian Foragers – Memories and Reflections*. http://radicalanthropologygroup.org/sites/default/files/journal/ra_journal_nov_2013_22-37.pdf.

———. (2014). *Anthropology, Ecology, and Anarchism – A Brian Morris Reader*. Oakland: PM Press.

Pandya, V. (2009). *In the Forest – Visual and Material Worlds of Andamanese History (1858–2006)*. Lanham; Plymouth: University Press of America.

———. (2010). Hostile Borders on Historical Landscapes – The Placeless Place of Andamanese Culture. In *The Jarawa Tribal Reserve Dossier – Cultural and Biological Diversities in the Andaman Islands*, edited by Pankaj Sekhsaria &Vishvajit Pandya. Paris: UNESCO: 18–29.

Pandya, V., & Mazumdar, M. (2012). Making Sense of the Andaman Islanders – Reflections on a New Conjuncture. *Economic and Political Weekly* 44: 51–58.

Perry, R. J. (1978). Radcliffe-Brown and Kropotkin – The Heritage of Anarchism in British Social Anthropology. *Kroeber Anthropological Society Papers* 51/52: 61–65.

Proudhon, P. (2011). *What is Property?* Hamburg: Tredition.

Radcliffe-Brown, A. R. (1922). *The Andaman Islanders*. London: Weidenfeld & Nicholson.

Raju, G. (2010). Ranchiwallahs – The Pains of Dispossession. *The Light of Andamans* 35: 2. https://in.groups.yahoo.com/neo/groups/andamanicobar/conversations/messages/6758.

Sahlins, M. D. (1974). *Stone Age Economics*. London: Tavistock Publications.

Scott, J. C. (1985). *Weapons of the Weak – Everyday Forms of Peasant Resistance*. New Haven: Yale University Press.

———. (2009). *The Art of Not Being governed – An Anarchist History of Upland South East Asia*. New Haven: Yale University Press.

Sen, S. (2000). *Disciplining Punishment – Colonialism and Convict Society in the Andaman Islands*. New Delhi: Oxford University Press.

———. (2009). Punishment on the Fringes – Maulana Thanesari in the Andaman Islands. In *Fringes of Empire – Peoples, Places, & Spaces in Colonial India*, edited by Sameetah Agha & Elizabeth Kolsky. New Delhi: Oxford University Press: 139–168.

———. (2010). *Savagery and Colonialism in the Indian Ocean – Power, Pleasure and the Andaman Islanders*. New York: Routledge.

———. (2016). Race, Aboriginality and the Adivasi: Some Implications for the Andaman Islanders. In *Manifestations of History – Time, Space, and Community in the Andaman Islands*, edited by Frank Heidemann & Philipp Zehmisch. Delhi: Primus: 74–95.

Sekhsaria, P. (2007). *Troubled Islands – Writings on the Indigenous Peoples and Environment of the Andaman & Nicobar Islands*. Pune: Kalpavriksh; LEAD-India.

———. (2009). When chanos chanos became Tsunami macchi – The post-December 2004 scenario in the Andaman and Nicobar Islands. *Journal of the Bombay Natural History Society* 106, no. 3: 256–262.

Sekhsaria, P., & Pandya, V. (Eds.). (2010). *The Jarawa Tribal Reserve Dossier – Cultural and Biological Diversities in the Andaman Islands*. Paris: UNESCO.

Springer, S. (2016). *The Anarchist Roots of Geography – Towards Spatial Emancipation*. Minneapolis: University of Minnesota.

Sundar, N. (2007). *Subalterns and Sovereigns – An Anthropological History of Bastar (1854–2006)* (Second Edition). New Delhi: Oxford University Press.

Vaidik, A. (2010). *Imperial Andamans – Colonial Encounter and Island History*. Hampshire; New York: Palgrave Macmillan.

Venkateswar, S. (2004). *Development and Ethnocide – Colonial Practices in the Andaman Islands.* Copenhagen: Iwgia.

White, R. J., Springer, S., & Lopes de Souza, M. (2016). *The Practice of Freedom: Anarchism, Geography and the Spirit of Revolt.* Lanham: Rowman & Littlefield.

Wolfe, P. (1999). *Settler Colonialism and the Transformation of Anthropology – The Politics and Poetics of an Ethnographic Event.* London: Cassel.

Zehmisch, P. (2016a). Undoing Subalternity? Anarchist Anthropology and the Dialectics of Participation and Autonomy. In *Negotiating Normativity: Postcolonial Appropriations, Contestations and Transformations,* edited by Nikita Dhawan, Elisabeth Fink, Johanna Leinius, & Rirhandu Mageza-Barthel. New York: Springer: 95–109.

———. (2016b). The Invisible Architects of Andaman: Manifestations of Aboriginal Migration from Ranchi. In *Manifestations of History: Time, Space and Community in the Andaman Islands,* edited by Frank Heidemann & Philipp Zehmisch. New Delhi: Primus: 122–138.

———. (2017a). *Mini-India – The Politics of Migration and Subalternity in the Andaman Islands.* New Delhi: Oxford University Press.

———. (2017b). Fluid Indigeneities in the Indian Ocean: A Small History of the State and Its Other. In *Indigeneity on the Move: Varying Manifestations of a Contested Concept,* edited by Nasir Uddin, Eva Gerharz, & Pradeep Chakkarath. Oxford; New York: Berghahn: 270–293.

———. (2017c). Anarchie auf den Andamanen? Ethnographische Reflexionen zum Spannungsfeld von autoritärer Staatlichkeit und Strategien der Herrschaftsvermeidung im Indischen Ozean. *Paideuma: Mitteilungen zur Kulturkunde* 63: 231–250.

Zubair, A. (2011). Once Upon a Time … http://lightofandamans.blogspot.de/2011_09_18_archive.html.

8

Blockading Hamburg

Green Syndicalism vs G20

Ryan Thomson

> *Law enforcement & military did a spectacular job in Hamburg. Everybody felt totally safe despite the anarchists.*
>
> – @PolizeiHamburg #G20Summit 2:50 PM
Jul 8, 2017 @realDonaldTrump

Smoke billowed across the horizon with police helicopters echoing overhead. As we neared the city centre, the metro came to a stop and a voice came over the speaker announcing it would no longer be running. Droves of people were already in the street proceeding in the same direction on bike or foot. The perimeter of the police-free zone was demarcated by successive barricades along the outskirts of St. Pauli, Sternschanze, and Schulterblatt neighbourhoods. Anti-G20 banners hung from nearly every window, balcony and rooftop. Despite vicious acts of state repression, border patrols, unlawful detentions and the larger siege, it was clear that the neighbourhoods had collectively defeated the police by maintaining their autonomy for the entirety of the Summit. On one side of the district, it felt like a block party with massive bonfires, music, free food and drinks. In other areas fighting escalated into a guerilla war between insurgents throwing with rocks and paramilitary soldiers armed with automatic rifles. Police did everything in their power short of using lethal ammunition, but it did little to slow the thousands of people flooding the streets to expand the confrontation, provide medical aid and organize jail support. Mainstream portrayals of the protests seldom speak to the intense comradery between strangers who collectively challenged the very legitimacy of global capitalisms' most powerful delegates.

The 2017 G20 Heads of State Summit took place in the Hamburg Messe convention centre bordering the radical boroughs which have been long recognized for their social centres, graffiti culture and the infamous football Ultras. De facto martial law had been declared banning freedom of assembly within 38 kilometres of the city. State security forces actively sought to prevent an influx of radicals by detaining people at airports and stopping entire trains at the border. In one instance, two full buses of young people were taken directly to a detention facility because the German police considered the passengers to be 'hoodlums'. Several locals were tied up while police raided their apartments for 'suspected scene affiliation'. There were numerous early reports of police driving into crowds and attacking people on their way home from work. These pre-Summit efforts to crack down on dissent set the stage for a historic confrontation between tens of thousands of anti-capitalist demonstrators and over 31,000 police.

For political ecologists, the G20 represents the highest realm of state policy; the political structure concerning all ecologies. Accordingly, the following chapter examines the Battle of Hamburg as a multi-scalar environmental conflict by building on Kropotkin's sociology of the state to define the Summit as a supra-state collaboration largely responsible for numerous acts of ecological ruination (Kropotkin 1896; Kinna 2015). In doing so, anarchist praxis is employed to explore political ecology concepts including Massey's event of place, Agrawal's environmental subjects and violence as habitus (Massey 2005; Agrawal 2005; Peet and Watts 2004). These concepts are then coupled with the anarchist ideas of green anarchism, anarchist geography, biopolitical civil war, direct action and temporary autonomous zones (TAZ) to establish a series of analytical bridges necessary for understanding the Battle of Hamburg. These four bridges seek to foster and sustain a long-overdue dialogue between political ecology and anarchist theory (cf. Death 2014). Take for example that both scholastic traditions have explored their own competing conceptions of liberation ecologies without acknowledging one another (Marshall 1998; Peet and Watts 2004). This divide offers an opportunity to bridge these fields and expand both in the process.

In hopes of informing present and future actions, this text demonstrates the contribution of anarchist political ecology by contextualizing the G20 Summit and corresponding conflict along a specific set of conceptual lines. The CrimethInc. Ex-Worker Collective has assembled a detailed account and analytical summary of the conflict in their zine *Don't Try to Break Us—We'll Explode* (2017). Unicorn Riot also provided in-depth coverage of events throughout the Summit despite sustaining serious injury. Building on their work, this chapter seeks to engage the idea of emancipation as an epistemological field which generated new collectivities throughout the streets of Hamburg. In doing so, I aim to describe the experiences of myself and my affinities throughout the Battle. Thus, the writing style is somewhat experimental, drawing on a combination of general portrayals, analytical concepts, quotes from friends and personal experiences.

The totality of the occupying force, referred to here as Fortress Hamburg, initiated and engaged in a violence that reached points comparable to that of war.

From my vantage point, the most serious acts of physical harm were carried out by the police and paramilitary deployments seeking to repress the dissent of tens of thousands of people. Throughout the Battle, many mainstream news agencies kept a running tab of property damages (which in the end totalled just over €12 million) without ever mentioning the costs of the two-day Summit (€130 million before reinforcements). They prioritized the minor injuries of the police (many of which were self-inflicted) while omitting the dozens of demonstrators hospitalized by state forces. With numerous people lying bloodied in the street, the hypocritical narrative of 'senseless property violence' grew increasingly absurd bordering on tragic. The myopic views and decontextualized explanations of mainstream media presented the conflict as existing entirely beyond reason. Even supportive vanguard Marxist and *red bloc* accounts failed to acknowledge the egalitarian ideals and logic behind the event (WSWS, LFI). And yet from the other side of the barricades, everything made complete sense. The tear gas presented what exactly was at stake with strikingly clarity; the Battle configured new collectivities and affirmed friendships. It constituted a moment of truth which questioned the unequal local, regional and global power relations as a unified system of coercion and domination.

KROPOTKIN'S SOCIOLOGY OF
THE STATE AND THE G20

If breaking windows and fighting back when the cops attack is 'violence', then give me a new word, a word a thousand times stronger, to use when the cops are beating non-resisting people into comas.

– Starhawk, After Genoa 2001

For Kropotkin, the state means something quite different from the idea of government. The state

> not only includes the existence of a power situated above society, but also of a territorial concentration as well as the concentration in the hands of a few of many functions in the life of societies... a whole mechanism of legislation and of policing has to be developed in order to subject some classes to the domination of others. (1896, 10)

The state is a particular form of social organization characterized by a hierarchical system and centralization (1922, 317f). The state and Capitalist property relations are indivisible pieces of the singular world-system.

> The state . . . and Capitalism are facts and conceptions which cannot be separate from each other. In the course of history these institutions have developed, supporting and reinforcing each other . . . they are connected with each other – not as mere accidental co-incidences. They are linked together by the links of cause and effect. (1995, 94)

He continues, 'The state has been, and still is, the main pillar and the creator, direct and indirect, of Capitalism and its powers over the masses' (1995, 97). To isolate economic processes from state power is both impossible and fails to consider the force required to maintain unequal property relations.

Domination and the assertion of monopoly power is the central function around which the state is defined and organized.

The state is an institution which was developed for the very purpose of establishing monopolies in favour of the slave and serf owners, the landed proprietors, canonic and laic, and merchant guilds and money-lenders, the kings, the military command-ers, the noblemen, and finally, in the nineteenth century, the industrial capitalist, whom the state supplied with 'hands' driven away from the land. (1903, 166)

In his historical review of the tragedy of the commons, he traces the elementary roots of the state to four central institutions.

This is the way the mutual alliance between the lord, the priest, the soldier and the judge, that we call the 'State', acted towards the peasants, in order to strip them of their last guarantee against extreme poverty and economic bondage. (32)

The heads of state and power elite are organized with the intention uphold monopoly as their central purpose.

This functional allegiance significantly differs from Max Weber's popular defini-tion whereby monopoly is viewed 'as an organizational feature accompanying the legitimate use of physical force within a given territory' (1919). Modern states, Kro-potkin argued, of course claimed a monopoly of the defence of territory, but their purpose was to serve as instruments for establishing monopolies in favour of the rul-ing minorities (Kianna 2015). Throughout his work, he invokes numerous examples of different types of monopolies including industry, production, education, religion, transport and even morality backed by law. The assertions of monopoly power over other populations is essential for understanding the dynamics governing the G20 member states.

[T]he state should be thought of not simply as a series of institutions and struc-tures of power, but as a certain authoritarian relationship, a particular way of think-ing and structuring our lives-and so the idea of a politics of autonomy from the state involves the development of alternative non-authoritarian relationships, political practices, ways of thinking and modes of living. (Newman 2010, 169–170)

These relationships, asserted by the state apparatus, are enforced in the name of monopolizing surplus extraction and seek to replace community relations based on mutual aid the world over.

Applying Kropotkin's sociology of the state to the G20 allows for the concept of monopoly to the industries and economic power structures to transcend the artificial borders of global capitalism. These broad monopolistic interests not only govern the G20's efforts to enforce their will externally on less powerful states but also inform the internal supra-state collaboration of monopolistic interests. Accord-ingly, the G20 Heads of State Summits have come to represent the grand political event of neoliberal capitalism and largely responsible for numerous acts of ecological ruination. In some instances, these types of have Summits taken place in remote areas such as the 2007 G8 in Heiligendamm's seaside resort (which served as a sort of comparative case for 2017 mobilizing) or the 2015 G7 in the Bavarian Schlooss Elmau Castle. On other occasions, the Summits have been held in large urban cities such as Pittsburg, Hangzhou and Hamburg.

Collectively, G20 countries presently account for 74 per cent of all greenhouse gas emissions and roughly 80 per cent of global gross domestic production. These twenty state representatives, their invitees and finance administrations are also many of the primary perpetrators of environmental injustice. Many of these acts of ecological and social destruction are occurring within the Global South. From Widodo's rampant deforestation of Indonesia's rainforests to Turnbull's violent seizure of indigenous land across western Australia, Zuma's exploitative diamond mining throughout South Africa, Temer's land war throughout Para Brazil and Macri's fisheries collapse off the coast of Argentina. The systemic violence of the G20 leaders in the Southern Hemisphere alone is too long to list. Even among more self-proclaimed leftists in the G20, Peña's recent paramilitary attacks on the Zapatista's and Xi Jinping's rush to industrialize China have resulted in the state criminalizing thousands of environmentalists. Similar destructive patterns can be found among centrist liberals such as Macron's brutal raids on Le ZAD (Zone of Autonomous Defense), Merkel's repression of the Hambach forest defenders and Abe's reintroduction of Antarctic whaling, which has pushed many marine species to the brink of extinction.

The new wave of authoritarians also participate in similar ecocidal behaviours. Modi's embrace of industrial farming has resulted in widespread water shortages, toxic run off into all major rivers and groundwater contamination in the name of advancing economic profit. Similarly, Putin's oil and gas development in the Arctic zone (AZRF) has resulted in massive contamination of the Arctic ocean and surrounding river basins. This has resulted in a massive whitefish die off, further undermining the livelihoods of approximately 250,000 indigenous across the region (Greenpeace 2012). Of the hundreds of solidarity gestures in the streets of Hamburg, the most frequent oppositional references were directed at dictator Erdoğan's bombing of Rojavan forces and their environmental infrastructure throughout greater Kurdistan (see *Strangers in a Tangled Wilderness* 2015).

The transnational dimension of the G20's monopolistic interests is also apparent even between liberals and the new wave of authoritarians. Perhaps the best example of this is Trudeau's Tar Sand mining expansion and Trump's use of the National Guard to ensure the 1,700-mile-long construction of TransCanada's Keystone XL pipeline across multiple indigenous reservations. Support for the Standing Rock Sioux, among other Mni Wiconi struggles, was also a popular topic for solidarity frequently expressed in the streets of Hamburg. These examples of violence and repression are seldom framed as such given a distinction between types of violence and the interests they seek to uphold.

In sync with Kroptokin's emphasis on monopoly expansion, Deleuze and Guattari describe how the state uses law to differentiate between types of violence (war, crime, policing and structural violence) enabling the claim of legitimacy (of capture) and presupposition of subjugation.

War . . . implies the mobilization and autonomization of a violence directed first and essentially against the State apparatus. Crime is something else, because it is a violence of illegality that consists in taking possession of something to which one has no 'right'. . . . But State policing or lawful violence is something else again, because it consists in capturing while simultaneously constituting a right to capture. It is an incorporated, structural violence distinct from every kind of direct violence. The State has often been defined by a 'monopoly of violence', but this definition leads back to another definition that describes the State as a 'state of Law' (Rechts-staat) . . . State or lawful violence always seems to presuppose itself, for it preexists its own use: the State can in this way say that violence is 'primal', that it is simply a natural phenomenon the responsibility for which does not lie with the State, which uses violence only against the violent, against 'criminals' . . . in order that peace may reign. (1980, 448)

This crucial differentiation allows for state representatives to emphasize the criminality of those opposing them while downplaying the vicious actions of state actors both domestically and abroad. These hypocritical dynamics are crucial for maintaining a system of structural violence and justifying brutal police actions in the streets. To differentiate types of violence along legal lines enabled the state to claim a wide variety of heinous actions rightful. In Hamburg, this ranged from beating a non-violent demonstrator into an artificial coma, 'portable courtrooms' overpacking impromptu prisons, to hundreds of other human rights violations.

'Property is a relation of domination and it should be understood as a distinctive form of violence' (Springer 2015). For Proudhon, these property relations are not only illegitimate but constituted an act of theft enforced through coercion and violence. As paraphrased by a friend from Hamburg, 'The same system that guards the banks of hoarded gold is the same system that enforces austerity while criminalizing the poor. All cops enforce inequality.' While I consider myself fundamentally opposed to violence, I refuse to engage in the legalistic differentiation that criminalizes acts of resistance and self-defence in the name of preserving the status quo. Accordingly, to smash a bank window or defy a ban on assemblies represents a pushback against violence. To throw a can of tear gas away from a crowd is an act of self-defence. These actions can hardly be considered violent because they maintain no impetus for coercion or domination but rather a desire for self-preservation. By criminalizing resistance efforts, the soldiers and the judges sought to insure the monopolistic interests of the G20 capitalists. Little did the authorities suspect that Hamburg locals would resist state occupation so vigorously. As articulated by a local friend, 'This is a Battle between Hamburg and neoliberalism's stormtroopers of capital.'

Environmental Subjects and Green Anarchism

Humanity is nature becoming self-conscious.

– Élisée Reclus 1905

Within political ecology, Agrawal's concept of environmental subjects is immediately relevant to describe 'those for whom the environment constitutes a critical domain of thought and action' (2005, 16). The concept investigates how actions, ideas and identities are entwined with the necessities and complexities of power imposed by an environmental regime. As described by Robbins (2012),

Correlatively, new environmental regimes and conditions have created opportunities or imperatives for local groups to secure and represent themselves politically . . . these relationships are emerging key themes, as are the potentially emancipatory implications of very different situations where people's demands for autonomy are linked to other environmental practices and political identities around the world. (216)

In emphasizing new types of political action, environmental subjectivity emphasizes the connection of various ecological strands across links between disparate groups, across class, ethnicity and gender.

The politicization of the climate and environmental issues received a central place within the *Solidarity without Borders* call to blockade the Summit (g20-demo.de) among numerous other demonstrations focused specifically on climate change and environmental issues. Ecology was a central theme at the Global Solidarity counter-Summit, where the keynote speaker was Vandana Shiva. The largest of these strictly environmentally focused protests was the first big demonstration the Sunday before the Summit. G20 Protestwelle (protest wave) brought thousands out to the Binnenalster (inner Alester lake), Rathausplatz and nearby Rathausmart City Hall in the old town quarter. The strictly green demonstrations were primarily coordinated by NGOs although numerous unions and even a few anarchist collectives were in attendance. The green thread was immediately apparent by the hundreds of signs, banners and speeches especially so at the Solidarity without Borders demonstration on the final Saturday.

The drawback of these protests was that they maintained the impression of what my friends referred to as 'bürgerlich' (civil discourse) in their legalistic approach that made moral appeals to those in power. Such efforts tend to legitimate state power and perpetuate an understanding of social change that reinforces the idea that the problems are the bad politics of specific G20 rulers and not the structure itself. In doing so, these types of groups accept the status quo as a given and limit themselves to a 'practical strategy' of working within the system. As anarchists supporting a diversity of tactics, this approach certainly had its place within the larger strategy and managed to voice green concerns that many liberals could engage. As the uprising unfolded, many of these groups also stepped up their actions, exemplified by twelve Greenpeace boats, which temporarily blockaded coal shipments and successfully rushed the harbour. This further scattered police efforts across the city's land, air and water, effectively adding another territory that required policing. It is worth noting that police ships were primarily used to deploy water cannons at demonstrators on land and received waves of projectiles pelting the ship in return.

Several demonstrations managed to bridge the divide by engaging both types of critiques and tactical approaches. Even while attending many of the symbolic demonstrations, it was tough to differentiate diverse political currents. Solidarity and a diversity of tactics were broadly embraced. The NoG20 mobilizing committee coordinated many of the larger demonstrations alongside a citywide social strike. Among these efforts, the Climate Action *Unplug G20* effort coordinated blockades on Hamburg Harbor to stop the importation of coal in solidarity with activists challenging extraction in Columbia.

[T]he destruction of natural livelihood is being tacitly accepted for the purpose of a successively economic growth. So we are now taking it into our own hands and intervene where the problems are being produced, for example in the logistics of global trade or in the utilization of fossil energy sources. (Fischer, G20-Protests Press Release)

While some organizations embraced bürgerlich strategies for political discourse, a far more radical spirit could be found just behind the barricades; green anarchism. There were several different green anarchist currents present among the participants mobilizing against the G20.

Over collective meals, between actions, and even while in the streets, I participated in a variety of ecological discussions ranging from climate justice, green syndicalism, veganism and insurrectionary social ecology. These ideas were not competing but rather drew upon one another as complementary strategies, a sort of green synthesis situated beneath a global total liberation effort (Pellow 2014). 'We sought to mobilize the greatest force possible, just like they did. So, we reached out to the unions, the collectives and affinity-groups, and even anti-oppression individualists; we all live on the same dying planet.' Many of the positions were represented symbolically, a black and green flag or a clever banner. Solidarity gestures to environmental struggles abroad were also commonplace.

Radical environmental struggles were not only apparent by the flags and banners of the demonstrators but in the graffiti left behind on city walls. A few examples include 'ZAD Partout' in the fish market, ecology symbols on the cracked glass of corporate banks and 'We are all Zapatistas' on the boarded-up windows of department stores. Many of the local struggles previously mentioned sent individuals who gave speeches to kick off the *Welcome to Hell* demonstration. As the battle unfolded, it was incredibly powerful to see images of solidarity actions sent from the frontlines abroad. These solidarity gestures, especially the actions and graffiti from Hambach and Rojava meant so much. Solidarity went beyond a simple invocation or sign once it was clearly reciprocated. NoG20 solidarity demonstrations and marches in Exerchia, Crete, Madrid, Kiev, Stockholm, Amsterdam, Vienna, Hanover, Berlin and Paris (among numerous other cities) only further this grand feeling.

It would be misrepresentative to present the Battle of Hamburg as a revolt of formal workplace unions and environmental groups demonstrating in the name of environmental defence. However, it would be equally difficult to engage the multifaceted mobilization in a vacuum that fails to recognize the contextual importance of climate change and the numerous local environmental conflicts. Many of the participants mobilized as members of radical unions while others participated as extensions of informal collectives and impromptu affinity groups. Either way, the outcome was a coming together. Green emphasis was intermixed among other currents including labour, feminism, queer and immigrant struggles.

THE ANARCHIST GEOGRAPHY OF
HAMBURG AND EVENT OF PLACE

The global dimensions of the G20 explain why the protests happen. It does not explain how the demonstrations managed to outmanoeuver state forces and win over local support. The city of Hamburg, with its rich history and vibrant autonomous scene, constitutes an anarchist geography; a terrain where radical praxis has the homefield advantage (Davidson 2014). Situated on the Elbe River, Hamburg is the second largest city in Germany and third largest Port City in Europe. The port, Germany's *Gateway to the World*, established the city as a hub for international migration and cultural diversity comparable to nearby Amsterdam, Dortmund and Rotterdam. This multiculturalism is further complemented by a sizeable population of young people. Hamburg is also a home to nineteen universities accommodating nearly 100,000 students, many of which participated in coordinating a variety of militant actions during the Summit. For example, student unions coordinated the Bildungsstreik (student strike) 'students against the learning factory', effectively shutting down all of Hamburg's universities. The student march with just over two thousand young people established a sizeable blockade along the port on Friday morning. Their march route and blockade demonstrated a strong of understanding of historical-geographic roots emerging three decades earlier.

While the AIDS epidemic ravaged the Reeperbahn, the recession of 1980 led to mass unemployment with many empty buildings falling into disrepair in the centre of the city. This provided an opportunity for communities of squatters looking to form new ways of communal living. The squats brought together a diverse community who created radical social centres: info-shops, bookstores, coffeehouses, meeting halls, bars, music venues, art galleries and other grassroot multiple-use spaces (Young 2001). Specifically, the initial Hafenstraße squats in St. Pauli initiated in 1981 became home to a vibrant autonomous scene that managed to fend off numerous police and fascist attacks. This fostered the development of the black bloc form of militancy that mitigated 'the problematic dichotomy between popularly executed non-violent civil disobedience and elite, secretive guerilla terrorism and sabotage' (Young 2001). Although the Hafenstraße has since become a housing cooperative, numerous spaces such as Onkel Ottos Bar have kept the counter culture going strong.

While there was no planned effort by anti-capitalists to infiltrate and radicalize the St. Pauli football club, a few key punks started attending matches. To the disdain of old left intellectuals, young radicals and their friends began attending St. Pauli football matches at the nearby Millerntor stadium throughout the 1980s. With time, the militant 'Hafenstraße Bloc' fans grew and brought with them emancipatory ideas as conveyed by their banners, songs and imagery (Davidson 2014). Supporters of the club embraced their working-class identity and adopted the Jolly Rodger skull and crossbones flag from the squats to symbolize the historical conflict between rich and poor. Ultra Sankt Pauli is now infamous for their fervour and radical politics. Football has in part been a big part of developing a communitarian sense of identity throughout the city.

Rote Flora, a music theatre squat started in 1988, located to the north in the Sternschanze quarter (referred to simply as *Schanze*) is now the primary autonomous centre of Hamburg (rote-flora.de). The centre served as an important location during the Summit, only a few blocks from the Hamburg Messe convention centre. In December 2013, when the government threatened to evict and demolish the centre, Rote Flora became a symbol of resistance to gentrification. This culminated in the December 21 riots, when more than 7,000 people clashed with police in solidarity with Rote Flora and Lampedusa, a political group of migrants in Hamburg. The autonomous scene remains strong to this day, thousands of radical stickers and graffiti cover the city like moss providing a visual reminder of present organizing efforts. In addition to Ultra Sankt Pauli and Lampendusa, Hamburg has multiple decentralized anti-fascist and anti-racists collectives active in different districts throughout the city (e.g., antifa 309). The communities of Hamburg have a cultural predisposition to autonomous resistance and decades of mobilizing experience on this terrain.

Massey's Event of Place

Political ecology emphasizes issues of spatial and temporal scale to analyse the unfolding of contention and conflict across space. On one hand, this acknowledges regional legacies of resistance and corresponding mobilization networks as the vehicle for political escalation. Conversely, this attention to scale also informs resistance to the G20 Summit by emphasizing the transnational relations to struggles abroad, and the interplay of power and scale across space. The G20's destruction decisions reach from the Unist'ot'en territory in Canada to the battlefields of Rojava and to the frontlines of Indonesia. Accordingly, G20 mobilizing efforts sought to vocalize their support and solidarity for these movements abroad.

This illustrates Massey's (2005) notion of event of place. ' "Here" is an intertwining of histories in which the spatiality of those histories (their then as well as their here) is inescapably entangled. The interconnections themselves are part of the construction of identity' (139). The local symbols, landmarks, mobilizing strategies and counterculture milieu are inseparable from the revolt against police repression. Regionally, prior experiences at the nearby G8 Summit in 2007 and even Blockupy in Berlin served as tactical assessments for what could be expected. These experiences also informed the mobilization strategies employed in 2017.

[W]hat is special about place is precisely that thrown-togetherness, the unavoidable challenge of negotiating a here-and-now (itself drawing on a history and a geography of thens and theres); and a negotiation which must take place within and between both human and nonhuman . . . is not just that old industries will die, that new ones may take their place. (Massey 2005, 140)

From the creation of the port to dock-strikes, to the autonomen squats and graffiti culture, to the rise of the St. Pauli Ultras and antifa movement chapters, history is a force to be reckoned with in the radical present. Hamburg's radical history served as a clear juxtaposition to the G20's shared financial and political interests.

BIOPOLITICAL CIVIL WAR AND STATE REPRESSION

The territorial question isn't the same for us as it is for the state. For us it's not about possessing the territory. Rather, it's a matter of increasing the density of communes, of circulation, and of solidarities that the territory becomes unreadable, opaque to all authority. We don't want to occupy the territory, we want to be the territory.

– The Invisible Committee 2011, 108

Massive police raids on Entenwerder, Wohlers, Altona, Königstrasse, Moorfleet and Gählerspark protest camps set the tone early on. Most of the camps were small, hosting under twenty tents. Following these early raids, non-activist institutions, such as the Schauspielhaus theatre began to step in offering space for people to sleep. Police began sporadically deploying water cannons against small gatherings throughout the city but for the most part kept their distance from the bürgerlich-style protests.

The crucial juncture came on 6 July, where long-time members of the autonomist scene had called for a massive rally titled 'Welcome to Hell'. It was a diverse anti-capitalist gathering at the St. Pauli fish market with free food, drinks, speeches and live music. When the programme concluded,

> Several sound trucks playing a variety of revolutionary music moved to the front of the crowd, followed by one affinity group after another, participants pulling black rain jackets and gloves over their colourful summer clothing. Line after line formed: this was the Welcome to Hell black bloc. (CrimethInc 2017)

Police allowed the permitted march to go a couple hundred metres before stopping the demonstration with a wall of riot cops and four water cannons, resulting in a tense stand-off. Negotiations looked to be developing when the police launched simultaneous attacks from multiple directions using snatch squads, tear gas, flash bombs and water cannons. The trap took many by surprise but roughly fifty individuals in the black bloc created a new formation opposite the police. A section of the bloc heroically held the line for about five minutes, allowing for hundreds of their comrades to scale the southern wall towards the river. It was immediately apparent the police sought to isolate the militant front section of the demonstration; it did not take long for the police to start targeting anyone in the vicinity. Several medics were assaulted when they attempted to reach the injured, numerous media correspondents were beaten and their recording devices destroyed.

The port quickly filled with smoke and tear gas as the 'Keep Global Trade Open' billboard displayed on a shipping freighter across the river fell out of view. Flashbang explosions, coughing, distant screaming and the robotic voice over the loud speaker made it difficult to assess exactly what was happening. Shoes and broken cameras were strewn about where the police first attacked. While police pursued sections of the crowd down different streets and ally-ways, spectators began to join in by raining down rocks and bottles on the police. A mixture of black bloc, demonstrators and enraged spectators formed new lines behind the water cannons and police lines preventing them from moving. As the crowds adapted to the situation, new opportunities began to present themselves. While the police were busy with the attack in the fish market, a couple hundred people regrouped and faced off with the police along the border of Arrivati Park.

Before the Summit even started, the police's attempt to assert control through brutal force had lost their last claim to legitimacy. Despite a massive press campaign by the state ahead of the event declaring protesters as 'violent'; their propaganda effort collapsed. The police attacks had turned the city against them and they were viewed as an occupational force. On a local scale, they were unable to recover the narrative from that point on. The night before the Summit, resistance actions spread quickly with small groups fanning out and constructing barricades throughout the nearby districts. The materials from the construction sites of gentrification were the most accessible for constructing barricades. The symbols of Fortress Hamburg were the primary targets; police cars, G20 diplomat convoys and other symbols of capitalism such as billboard advertisements and ATM machines. All around the city, banks, chain stores and luxury cars were attacked throughout the night. These actions emanated outward from harbour and quickly spread as far north as Osterstrasse where several shop windows were broken.

As things began to slow along the harbour, it did not take long for anti-capitalist chants to turn small crowds into a sizeable gathering. Two demonstrations reformed a few blocks away, it took about an hour for these crowds to find one another. The police managed to prevent the crowd of roughly 12,000 people from continuing to march north for roughly an hour. We made it a couple miles before the police attacked a second time with a much weaker force: two water cannons and flashbangs. The fighting intensified along the larger intersections with large crowds seeking to avoid getting trapped in the smaller side streets. The police were spread thin and began intensifying the use of water cannons and snatch squads in hopes of disbanding the crowds. This move only further hastened frequency of street clashes and further emboldened resistance efforts.

Early the following morning targeted 'Color the Red Zone' blockades of key intersections further disrupted the Summit. Each of the blockades had a different colour and was surprisingly successful. The colourful blockades were juxtaposed by Hamburg Police pleading with locals to stop supporting the black bloc and revoking official press accreditation to limit the counter-narrative. In a sort of act of vengeance, police 'evidence collection' units began raiding social centres. LKA (police authority of the federal state) raided the international centre in the Brigittenstrasse St. Pauli, a nearby movie theatre, private homes and food coops. The raids were unable to find the conspiratorial evidence but resulted in at least six individuals being seriously injured. We began hearing news of different raids as the 10k *Critical Mass Bike Ride* toured the district.

The deployment and relative mobility of repressive technologies were stunning. A fleet of over two dozen of 'state-of-the-art WaWe 10,000' water cannons served as a sort of support vehicles. They frequently rotated deployments and were reliant on fire-hydrants to refuel. After the first night of autonomous resistance, it was apparent the police and their resources were clearly overwhelmed. The police were forced to call for an additional 10,000 reinforcements. The increased numbers were unable to quell the demonstrations and decentralized black bloc actions the following day. On the first night of the Summit, large barricades were set ablaze throughout Saint Pauli and Sternschanze to prevent police from gaining access. Flares, home-made fireworks and lasers were used to defend barricades along key intersections.

There were two strategies for protecting windows during the Summit. Some stores boarded up windows with plywood and metal wire grates. Such a strategy, employed almost exclusively by corporate chain stores, was highly ineffective. These corporate stores stuck out and, in many ways, made themselves targets. The more effective strategy, commonly adopted by local businesses, was to put up a sign saying *NoG20* or *Fuck Trump*. Stores with these signs in their windows went completely untouched. Corporate stores that failed to put up a sign denouncing the G20 were sacked across the street from local shops which were left untouched. These were clearly acts of differentiated property destruction; they occurred with a clear pattern and purpose. The pressurized containers obtained from the stores were thrown into the barricade fires creating loud explosions which echoed throughout the districts. No buildings were set on fire.

Several SEK (and Hamburg's local MEK) 'Cobra' Civil War and GSG9 Special Deployment Commandos units were activated just before midnight. These para-military forces were armed with assault rifles and dressed in camo. Their militarized invasion of Schanzenviertel was initially unsuccessful as thousands converged on their deployment. SEK threatened to shoot several of the medics who remained in the streets. They finally reached Rote Flora sometime after 3 a.m. By that point, most of the sizable crowd had disappeared into the night.

The Battle continued until Saturday morning with things quieting down shortly before the *Solidarity without Borders* demonstration. Despite the massive repression efforts, the demonstration drew over 80,000 people. The demonstration was led by the 'Kurdish Bloc' that prevented police from targeting the anarchist contingents. Police wearing black balaclavas lined the streets and sporadically attacked the protest which stretched well over 2 kilometres in length. The demonstration concluded with a concert across the street from Millerntor stadium. Many contingents were still enraged over police kidnapping their friends and continued fighting elsewhere throughout the district. During that final afternoon, barricades once again were established as the G20 Heads of State attended the Elbphilharmonie concert. Both events were live-streamed alongside one another contrasting two competing conceptions of power to the tune of Beethoven's *Ode de Joy*. The bored faces of Merkle, Modi, Trump, Widodo and others were presented immediately alongside police water cannons clashing with demonstrators a few blocks away.

In the months since the Battle, the German police state has initiated a new campaign of repression and mass manhunt they have titled 'Soko Schwarzer Block' (Hamburg Special Black Block Police Commission). This has included raids on dozens of homes and community centres, banning indymedia.org, social media surveillance, abducting and interrogating court observants without a warrant, publicizing hundreds of open criminal investigations and publishing photographs of a hundred people who participated in G20 resistance. The raids have been justified via claims of *Landfriedensbruch*, breach of the public peace. Right-wing judges in Hamburg have entirely discarded the pretext of criminal prosecution in their attempt to make a political statement against G20 prisoners by going well beyond the persecutors' recommendations. The investigation of over a hundred police officers for acts of misconduct has been swept aside for the time being. Two months after the Battle, the first trial sentenced a young detainee to two years and seven months for throwing a bottle and resisting arrest by curling into a foetal position while being hit with batons. United We Stand, a regional solidarity campaign against G20 repression, continues to organize support for our friends still being held captive.

Biopolitical Civil War

Fortress Hamburg with all its organization, technology, tactics, surveillance, propaganda and manoeuvring embodies an experiment in state repression. Although the apparatus of control failed to prevent the creation of a TAZ, the military occupation offers a case study in civil war governance. For Foucault, civil war is an application of biopower, 'The extension of state power over both the physical and political bodies of a population' (46, 2008). This vast and wide-ranging technology of control manages populations to the point where its dictates are internalized (Campbell and Sitze 2003). More than a disciplinary mechanism, biopolitics acts as a control apparatus exerted over entire populations. The practice of modern nation states and regulation of their subjects occur through diverse techniques aimed at subduing bodies. This system seeks to limit the creation of new collectivities by employing everything from misinformation, transportation policing and threats, in addition to brutal physical repression.

> Civil War is the matrix of all power struggles, of all the power strategies, and consequently, the matrix of all the struggles over and against power. Civil war not only brings collective elements into play, but it constitutes them. Far from being the process through which one comes down again from the republic to individuality, from the sovereign to the state of nature, from the collective order to the war of all against all, civil war is the process through and by which a certain number of new collectivities that had not seen the light of day constituted themselves. (Foucault 67, 2006)

It is only after the failure of other strategies that hundreds of people were captured, shipped miles south of the Elbe river to state facilities, locked into overcrowded cells in a blistering warehouse (which had previously been a wholesale food market), refused legal support and provided minimal food.

This biopolitical reasoning has been advanced by Tiqqun (2010) whose definition does not highlight expressions of class antagonisms but rather the persistent continuation of civil war through subjugation and depoliticalization. Their definition expresses how the G20 seeks to manage the omnipresent challenges to their claims of normality and its corresponding economic property relations.

> Empire perceives civil war neither as an affront to its majesty nor as a challenge to its omnipotence, but simply as a risk. This explains the preventive counter-revolution that Empire continues to wage against anyone who might puncture holes in the biopolitical continuum. Unlike the modern State, Empire does not deny the existence of civil war. Instead, it manages it. By admitting the existence of civil war, Empire furnishes itself with certain convenient means to steer or contain it. (Tiqqun 2010, 58)

These new collectivities took many forms during the Summit; the sum of which managed to puncture a hole in the biopolitical continuum creating a temporary autonomous space.

DIRECT ACTION, AUTONOMOUS SPACE AND VIOLENCE AS HABITUS

I always secretly looked forward to nothing going as planned. That way, I wasn't limited by my imagination. That way, anything can, and always did, happen.

– CrimethInc

Many anarchists prioritize direct action alongside other principles such as decentralization, autonomy, direct democracy, free associations, bioregionalism and mutual aid solidarity. Voltairine De Cleyre describes these types of activities as bold cooperative assertions of self-empowerment that can take many forms (1912). For De Cleyre, direct action can be classified into positive and defensive forms. Both types were present in Hamburg: the City Strike and communal kitchens were examples of positive acts and throwing back tear gas was defensive.

The principle of direct action provides the means for directly taking power by rejecting the heads of state who claim to represent their interests; it offers an alternative to the states electioneering, representative voting and legislation. While many syndicalists emphasize direct action in the workplace, 'emancipation through practical action' can occur anywhere (Bakunin 1869). In the words of Emma Goldman, '[d]irect action against the authority in the shop, direct action against the authority of the law, direct action against the invasive, meddlesome authority of our moral code, is the logical, consistent method of Anarchism' (1910, 62–63). Building on this broad definition, Rudolf Rocker summarized the tact as 'every method of immediate warfare by the workers [or other sections of society] against their economic and political oppressors' (Anarcho-syndicalism 1938, 78). The emphasis being placed on immediacy of resistance to hierarchical subjugation. Political ecologists have engaged direct action with, relative to many scholastic anarchists, emphasized the concept of agency and narrative development as a driver of grassroots praxis (Heynen and Sant 2015). Within sociology, direct action politics tends to be associated with and viewed in terms of a contentious performances framework (see Tilly 2008).

While the Summit exhibited many popular forms of direct action, from massive 'refugees welcome' (*Alles Allen*) dance parties, to looting and targeted blockades, as well as a few tactics that were new to me. 'Cornering' was commonplace, small gatherings of people scattered across the city. The strategy sought to undermine the ban on assemblies and was for the most part a creative success. 'I had expected the rage of activists. But I found the really angry people on the sidewalk, grilling' (Crimethinc 2017, 17). As I walked through the Schanze neighbourhood with friends, a few locals offered us drinks and plates of food as they sat outside their flat. The neighbours had pointed their speakers out of the window. The police sent two groups of cops to watch us, down the street we could see a handful of other groups doing the same thing. In the context of the Summit, hanging out in the street became a small gesture of resistance. Most of the discussions that have informed this text occurred while cornering. Fortress Hamburg turned even the smallest acts of determination into grand gestures of negation.

These defiant instances have transformative potential for its participants but also demonstrate the presence of oppression and effectiveness of resistance tactics to spectators. 'Direct action is always the clamorer, the initiator, through which the great sum of indifferentists become aware that oppression is getting intolerable' (De Cleyre 1912, 231). The attack on the Welcome to Hell march and liberation of Schanze spread discontent throughout the entire city. Even to areas not known for radical organizing, locals pulled their own trashcans into the street, leading to police spreading themselves thinner over more territory. Fortress Hamburg, overrun with hundreds of simultaneous direct actions, was unable to prevent the creation of a TAZ. The TAZ concept was introduced by Hakim Bey to describe the sociopolitical tactic of creating momentary spaces that elude formal structures of control in an era of closed maps (1991). Spatial emancipation, in the form of a police-free enclave, emerged throughout Hamburg as a sort of festival against repression which managed to pierce the grid of social control. Prior to the Summit, the creation of a TAZ felt incredibly unlikely given the sheer scale of psychological and physical repression. And yet, the shared ideas of a critical mass manifested as a riot.

The Schanze was autonomous in the sense that the mass of people, based on principles of self-directed and free associations, pushed the heavily militarized police forces out of the district. In Bey's terms, Fortress Hamburg was confronted with an encampment of thousands of guerilla ontologists. The intense event gave shape and meaning to a grand collective experience; an irreversible integration and affirmation of friendship unlike anything I have ever experienced before. Representations of the demonstrations seldom speak to the intense comradery which bonded strangers. These bonds physically challenged state forces and contested the legitimacy of the state's corresponding neoliberal world view in doing so. The words of G20 prisoner Riccardo Lupano illustrate ontological anarchy's call to create our own day.

> In the streets of Hamburg, I have breathed uncontrolled freedom, active solidarity, the determination to reject the deadly order imposed on us by the rich and powerful. No endless series of cars and orchestrated processions that cement the oppressive, murderous liturgy of capitalist everyday life. No blurry masses, forced to sweat and hump for the wealth of a disgusting boss. No thousands of absent pairs of eyes, directed at any aseptic display that distorts and alienates our experience of everyday life. I saw individuals looking into the sky and trying to grab it. I saw women and men who gave shape to their creativity and their oppressed dreams. I saw the energy of each one trying to reach out to others and not to rise above others. I saw the sweat on the forehead of those who sought their own desires, and not those of their tormentors.

By refusing state forces to opportunity to possess the city, a message was sent to capitalism's most powerful actors.

The territorial logic of the demonstrations made the TAZ a worthy tactical end but was it was not without its limitations.

We're not touting the TAZ as an exclusive end in itself, replacing all other forms of organization, tactics and goals. We recommend it because it can provide the quality of enhancement associated with the uprising without necessarily leading to violence and martyrdom . . . which liberates an area (of land, of time, of imagination) and then dissolves itself to re-form elsewhere/elsewhen, before the state can crush it. (Bey 1991, 2)

The dispersal of Hamburg's TAZ eluded formal structures of hierarchy and control. The conclusion of the Summit redirected the collective effort to local prisoner support and mobilizing to defend the Rote Flora from criminalization by Mayor Scholz. The Sunday after the Summit, a sizeable *Nobody Forgotten, Nothing Forgiven* demonstration to the south of the city was coordinated outside the barbwire of the Neuland impromptu prison before marching on nearby Harburg Townhall. Prisoner support has remained a regional issue ever since, perhaps more so now as a few trials are underway.

Violence as Habitus

Within the context of the Battle of Hamburg, direct action in the form of a TAZ teetered on what political ecologists refer to as violence as habitus. Drawing on Bourdieu et al. (2004) coined the concept to emphasize the discursive and institutional field of struggle where violence is normalized to the point it becomes habitus. Peet and Watts contend this field is driven by environmental processes which alter the systems of physical and symbolic violence. Consistent with Bourdieu's theory of symbolic power, political ecologists stress the epistemological roots of the uprising emphasizing the shared subjective experiences of its participants (1989). The schemes present in Hamburg were not mere habits, rather these shared dispositions were constituted as attitudes, mannerisms, tastes, moral intuitions and habits. These schemes allowed for the search for new solutions without calculated deliberation; they were both structured by past social positions and the structuring of future life. This both differs from and complements Bey's idea of ontological anarchy associated with the TAZ to challenge capitalist systems of knowledge (1991). An autonomous geographic imagination, whether temporary or permanent, is a big part of spreading violence as habitus.

CRITIQUE OF THE BATTLE

Praxis requires a critical reflection on strategy, implementation and application of ideals. The statement issued by the Free Workers' Union in Hamburg (FAU) regarding the G20 offers one notable critique of how direct action was realized. The local FAU chapter decided not to join in the call for protests as a union but instead participated as individuals while opening their social centre (the Black Cat) as a refuge for physical and emotional aid. Their statement, released over a month before the Summit, raises the question: Was the Battle of Hamburg an example of positive direct action? They contend that no, the demonstrations did not present an opportunity to improve the quality of life for their members. 'At best – with a high personal risk and huge financial expense for those calling for protests – the Summit will be interrupted and continued elsewhere.' For many of them, the Battle was at best a symbolic and defensive victory largely removed from the local struggles. Their point is certainly valid and echoes some of the stronger criticisms of Summit hopping. While resting after a meal at the Black Cat, a friend pointed out that the G20 presented

> a great opportunity to materialize transnational solidarities but ultimately lacked a positive offensive strategy for undermining the state's numerous systems of power. As a result, such events once again place us on the defensive . . . symbolic victories, even on a global scale, can only go so far.

We spent the rest of the meal discussing the effectiveness of different mutual aid strategies and current challenges facing local refugees' welcome organizing efforts.

CONCLUSION

The free city of the future does not belong to any nation.

                                    – From an Anti-G20 Banner in Liberated Schanze

The Battle of Hamburg unfolded with a clear logic; an epistemological clarity manifested as new friendships and a steady escalation of direct action tactics. This chapter has studied the event to illustrate a series of bridges between anarchism and political ecology as grounds for studying the conflict surrounding the G20 Head of State

Summit. In placing primary emphasis on the monopolistic assertion of power, Kropotkin's sociology of the state, supported by Deleuze and Guattari's legal differentiation of violence and Newman's state relations, articulates a relational functionality that goes well beyond Weberian organizational coincidence. The G20, as a suprastate collaboration that perpetuates multi-scalar environmental conflict, illustrates the strengths of political ecology's interest. Anarchist philosophy is poised to benefit from this emphasis on geopolitical scales and their implications for transnational solidarity. Even in a world freed from state borders, scale remains important. In a reciprocal manner, anarchist critiques of power offer political ecology a participatory exploration of state violence and resistance. Put differently, anarchism provides political ecology a means to engage in praxis.

Agrawal's environmental subjects speak to the unification of transnational environmental regimes and their implications for the creation of autonomous identities. From Rojava to Unist'ot'en territory, these various identities were prioritized among the synthesis of green anarchist currents present in the streets of Hamburg. The port community reflects the rich local history: squatted spaces, organizations such as antifa 309, Lampedusa and Ultra Sankt Pauli, foster a vibrant autonomous scene among a relatively young and diverse demographic. The anarchist geography of Hamburg, when faced with the G20's biopolitical civil war governance strategies, illustrates Massey's event of place. The history of the neighbourhoods, Millerntor stadium, Rote Flora and Hamburg Messe cannot be separated from the local opposition to the G20, city strike and TAZ. Biopolitical control established the necessary conditions; it included the border policing, disinformation campaigns, camp evictions, mass detentions and attack on the *Welcome to Hell* march sparking the Battle.

The numerous solidarity references to conflicts against the G20's global environmental regime speak to a mobilizing strategy informed by an awareness of political ecology issues. Peet and Watts, through violence as habitus, would likely emphasize the epistemological challenge to the G20's neoliberal ideas and collapse of the police's claims to legitimacy among locals. Bey (1991) would likely point out how direct action manifested a TAZ and introduced their own truth in doing so. The bürgerlich demonstrations, counter-Summit, as well as more radical *Color the Red Zone* and *Solidarity without Borders* blockades provided a tactical bridge for direct action to emerge as a justified response to state repression. The intersection of autonomies across land, time and imagination affirmed new friendships which made all of this possible. Although the Battle of Hamburg was not without its limitations, new offensive and positive forms of direct action are needed to move beyond symbolic victories. G20 prisoner Riccardo Lupano, who was recently sentenced to twenty-one months, integrates a series of points which speak to anarchist political ecology in his recent letter from Billwerder jail.

I am proud and happy to have been in the revolt against the G20 in Hamburg. The joy of the personal experience of the coming together of so many people of all ages and from all over the world, who have not yet subjected themselves to the total logic of money and the capitalist world, . . . the revolt showed a constant oscillation between internal war (special laws, border closures, deportations) and external war (massacres, destruction and poisoning of the planet Earth). Against the G20 . . . the possibility of people realizing freedom.

REFERENCES

Bakunin, M. (1869). *L'Égalité*, 14 August 1869.

Bey, H. (1991). *TAZ: Temporary Autonomous Zone, Ontological Anarchy, Poetic Terrorism.* Autonomedia.

Bourdieu, P. (1989). Social Space and Symbolic Power. *Sociological Theory* 7(1): 14–25.

Campbell, T., & Sitze, A. (Eds.). (2003). *Biopolitics: A Reader.* Duke University Press.

Crimethinc Ex-Worker Collective. (2017). *Don't Try To Break Us—We'll Explode: The 2017 G20 and the Battle of Hamburg: A Full Account and Analysis.* https://crimethinc.com/2017/08/08/total-policing-total-defiance-the-2017-g20-and-the-battle-of-hamburg-a-full-account-and-analysis.

Davidson, N. (2014). *Pirates, Punks, and Politics – FC St. Pauli: Falling in Love with a Radical Football Club.* Sports Books.

De Cleyre, V., & Berkman, A (Eds.). (2016). *The Selected Works of Voltairine De Cleyre: Poems, Essays, Sketches and Stories, 1885–1911.* Baltimore: AK Press.

Deleuze, G., & Guattari, F. (1987). *A Thousand Plateaus: Capitalism and Schizophrenia.* Minneapolis, MN: Minnesota University Press.

Fisher, J. (April 2017). *Press Release: G20-Protests: Disobedient Mass Actions Also in the Port of Hamburg.* https://shutdown-hamburg.org/index.php/2017/04/09/pressemitteilung-g20-proteste-massenhafter-ungehorsam-auch-im-hamburger-hafen/?lang=en.

Foucault, M. (2006). *The Will to Knowledge.* London, UK: Penguin Books.

———. (2008). *The Birth of Biopolitics: Lectures at the Collège de France, 1978–1979 (Lectures at the College de France).* Translated by Graham Burchell. New York: Palgrave Macmillan.

Goldman, E. (1910/1996). *Red Emma Speaks: An Emma Goldman Reader.* Amherst, NY: Humanity Books.

Greenpeace. (2012). *Oil and Gas Development in the Arctic: At What Cost?* Amsterdam, NL: International Report.

Heynen, N., & Sant L.V. (2015). Political Ecologies of Activism and Direct-Action Politics. In *The Handbook of Political Ecology.* London: SAGE: 169–178.

Kinna, R. (2015). Kropotkin's Theory of the State: A Transnational Approach. In *Reassessing the Transnational Turn Scales of Analysis in Anarchist and Syndicalist Studies*, edited by C. Bantman and B. Altena. London: Routledge: 43–61.

Kropotkin, P. (1896). *The State: Its Historical Role.* London, UK: Freedom Press.

———. (1903). *Modern Science and Anarchism.* London, UK: Freedom Press.

———. (1922). *Ethics: Origin and Development.* London, UK: George G. Harrap and Co.

———. (1995). *Evolution and Environment.* Montreal, CA: Black Rose Books.

McAdam, D., Tarrow S., & Tilly C. (2000). *Dynamics of Contention.* Cambridge, MA: Cambridge University Press.

McKay, I. (2008). *An Anarchist FAQ: Volume 1.* Oakland, CA: AK Press.

———. (2012). *An Anarchist FAQ: Volume 2.* Oakland, CA: AK Press.

———. (Ed.). (2014). *Direct Struggle Against Capital: A Peter Kropotkin Anthology.* Oakland: AK Press.

Marshall, P. (1998). *Riding the Wind: Liberation Ecology for a New Era*. London: Continuum Books.

Massey, D. (2005). *For Space*. London: Sage Publications.

Newman, S. (2010). *The Politics of Post-Anarchism*. Edinburgh: Edinburgh University Press.

Pellow, D. (2014). *Total Liberation: The Power and Promise of the Animal Rights and the Radical Earth Movement*. Minneapolis: University of Minnesota Press.

Robbins, P. (2012). *Political Ecology: A Critical Introduction*. Malden: Wiley Blackwell.

Rocker, R. (1938/2004). *Anarcho-Syndicalism: Theory and Practice*. Working Classics Series. Oakland, CA: AK Press.

Springer, S. (2015). *Violent Neoliberalism: Development, Discourse, and Dispossession in Cambodia*. New York: Palgrave McMillian.

———. (2016). *The Anarchist Roots of Geography: Toward Spatial Emancipation*. Minneapolis: University of Minnesota Press.

Springer, S., Birch, K., & MacLeavy, J. (2016). *The Handbook of Neoliberalism*. New York, NY: Routledge Press.

Strangers in a Tangled Wilderness (Ed.). (2015). *A Small Key Can Open a Large Door: The Rojava Revolution*. London: Combustion Books.

Tilly, C. (2008). *Contentious Performances*. Cambridge: Cambridge University Press.

Tiqqun (2010). *Introduction to Civil War*. Semiotext(e). Cambridge, MA: Intervention Series.

Watts, M., & Peet R. (Eds.). (2004). *Liberation Ecologies: Environment, Development, Social Movements* (2nd ed.). London: Routledge.

Weber, M. (1919/1946). *Politics as a Vocation*. Trans. and Ed. by H. H. Gerth and C. Wright Mills. New York: Free Press.

White, R. J., Springer, S., & de Souza, M. L. (Eds.). (2016). *The Practice of Freedom: Anarchism, Geography and the Spirit of Revolt*. London: Roman & Littlefield.

Young, D. D. (2001). *Autonomia and the Origin of the Black Bloc*. A-Infos News Service. http://www.ainfos.ca/01/jun/ainfos00170.html.

9

Rising above the Thinking behind Climate Change

World Ecology and Worker's Control

Ben Debney

We cannot solve our problems with the same thinking we used to create them.

– Albert Einstein

In even attempting to enter discussion around global warming, we are immediately confronted with multiple contending arguments and perspectives. At the baseline, we can generally agree that global warming exists, presenting the actual existential threat to human civilization that terrorism is alleged to as an article of moral panic (Debney 2017a). From there, we are confronted with multiple challenges – first, to identify the root causes and, second, to find solutions that, in addressing the root causes of global warming, reflect the truism articulated by Albert Einstein that 'we cannot solve our problems with the same thinking we used to create them'.[1] To do otherwise would be to reproduce the destructive dynamics we oppose through inadequate assessments of the root causes and the ineffective responses that derive from them, but which some find preferable perhaps due to their complicity with them. As a matter of characterization, trying to solve global warming with the thinking that created it is the best way to guarantee that the problem worsens, ever diminishing our chances of avoiding worst-case scenarios.

It is evident from many contemporary analyses of global warming that traces of the thinking that created it are apparent, both in terms of failing to address its root causes adequately, and in terms of failing to propose minimally adequate responses. First, liberal treatments of climate change markedly fail to distinguish between nature in the abstract and historical nature, between historical facts and a priori assumptions of liberal idealism. Second, where socialist treatments of climate change manage to transcend this shortcoming rather than reproducing it in the name of socialism, many still propose counter-strategies reflecting similar failures to distinguish between historical fact and the ideological baggage associated with Leninism, the 'authoritarian' end of the collectivist spectrum on the political compass.[2] While the thinking that produced the problem of climate change is mostly avoided, the thinking that produces the failure of the radical left to overcome capitalism is not.

Parallel to these conditions, we find in libertarian socialism ideas and principles that do manage a critique of the historical and social forces that give rise to the climate crisis, and which are at the same moment relatively free of the historical baggage that reproduces the thinking responsible for global warming insofar as they avoid the authoritarian prejudices and alienated social relations that perpetuate capitalism through the state. At the same time, the 'movement' broadly associated with libertarian socialism is by no means immune from problems, not least of which is the tendency to revert to alienated roles of permanent protest within radical ghettos of ideological purity, cut off from the community and the working class, where doctrinaire correctness takes precedence over having any social influence or capacity to positively influence our surroundings. Falling back onto individualistic solutions for collective problems, libertarian socialist theory and practice tend to suffer, failing then to link its basic theoretical concepts to contemporary realities, and in so doing, failing also to point the way out of the chaos, injustice and insanity of global corporatism.

Indeed, of all the possible criticisms of anarchism, one of the more prescient, arguably, is that it tends to be defined more by its opposition to coercive hierarchy rather than a positive vision of a baseline sane and just society, this choice of focus tends to reinforce the abovementioned alienated roles of permanent protest and ghettos of ideological purity. Unable to strategize beyond their ghettos and alienated roles, this criticism suggests that anarchists help to entrench and perpetuate the status quo by giving it the appearance of openness and plurality, while the far-right makes hay of the failure of the left to address the awesome destructiveness of global corporate supremacism among working-class communities. In addition to needlessly ceding ground to reactionary opportunists, it also suggests that our failure to fight for positive visions of alternate futures, positive visions that anyone concerned with a constructive outcome to present crises can use to take the initiative, means that we perpetually react to events instead of making them.

If Einstein's maxim follows, then it also follows that every problem, adequately grasped, contains the seeds of its own solution; to transcend the thinking that created the problem is to come to terms with the paradigm that produced that thinking in turn, and thus to be empowered in choosing different ways of thinking, being and acting, and so in both taking the initiative and bringing the fight to the powers that control and bore us. With a view to expediting this process, this chapter will examine responses to the climate crisis in the context of this truism of Einstein's, focusing on the extent to which it speaks to the tendency to perpetuate climate crisis through quick fixes that treat symptoms rather than causes in defense of privilege from change.

To that end, this chapter takes up the arguments made by World Ecologist Jason W. Moore, which examines the thinking underlying the social relations responsible for global warming by way of a critique of the 'society vs. nature' binary as an enabling ideological pretext, taking both as the basis for his alternative concept of the oikeios – a 'way of naming the creative, historical and dialectical relation in, between and also always within, human and extra human natures' (Moore 2015, 35; 2016; 2017). The oikeios is, in other words, the proverbial 'Web of Life'. As this chapter will argue, the concept of the oikeios provides an opportunity to expand on revolutionary praxis given its commonalities with traditional libertarian socialist notions of workers' self-management of production, providing an anchor upon which to meaningfully and effectively address the issue of overcoming the tendency to reproduce the thinking associated with the climate crisis in responding to it at the same moment.

THE CLIMATE CRISIS AND ITS THINKING

To date the more forward-thinking climate crisis analysis along the liberal spectrum has tended to reflect the insights of atmospheric chemist Paul Crutzen, who, in assessing the environmental changes associated with global warming, argues cogently that such changes are so profound as to constitute a new geological epoch. This he proposes to call the Anthropocene, or the epoch of man (Crutzen & Stoermer 2000). Crutzen locates the beginnings of the Anthropocene to around the time of the Industrial Revolution, while others, like eco-socialist John Bellamy Foster, trace its origins even later, to the initial period of nuclear weapons testing in the 1940s and 1950s (Angus 2016, 9). Considering the 'many other major and still growing impacts of human activities on earth and atmosphere, and at all, including global, scales', Cruzten wrote,

> it seems to us more than appropriate to emphasize the central role of mankind in geology and ecology by proposing to use the term 'anthropocene' for the current geological epoch. The impacts of current human activities will continue over long periods. According to a study by Berger and Loutre, because of the anthropogenic emissions of CO_2, climate may depart significantly from natural behaviour over the next 50,000 years. To assign a more specific date to the onset of the 'anthropocene' seems somewhat arbitrary, but we propose the latter part of the 18th century . . . during the past two centuries, the global effects of human activities have become clearly noticeable. This is the period when data retrieved from glacial ice cores show the beginning of a growth in the atmospheric concentrations of several 'greenhouse gases', in particular CO_2 and CH_4. Such a starting date also coincides with James Watt's invention of the steam engine in 1784. About at that time, biotic assemblages in most lakes began to show large changes. (Crutzen & Stoermer 2000)

The great value of Crutzen's insights is reflected in the scientific evidence indicating overwhelmingly that the environmental changes associated with global warming are profound and continuing (IPCC 2008). Nevertheless, a critical problem arises with the Anthropocene concept insofar as it attributes to humans per se, or nature in the abstract, what is a product of the prevailing mode of production, or historical nature (Cox 2015; Debney 2017b).

In positing human society and industry per se against nature, the Anthropocene idea inadvertently falls in with a binary logic that presupposes a split between society and nature where none is demonstrated (Moore 2015). In the beginning there was nature, says this binary, then human societies came along with their coal, gas and oil, and made a big old mess. This binary thinking identifiable in the root causes of the climate crisis itself – in terms, not least, of the Othering of those who got in the way of early capital formation (Said 1979; Davie et al. 1993; Deckard 2009; Williams 2012; Runehov et al. 2013). Its presence in critical commentary on the climate crisis is exactly what we mean when we talk about Einstein's truism about not being able to solve our problems with the same thinking we used to create them, especially where this leads to reinforcing the basic assumptions driving the problem and reproducing the conditions that allow it to fester.

PRIMITIVE ACCUMULATION

In contrast to the a priori myth of abstract nature in which the destruction of the natural environment is identified with the rise of human societies sans further elaboration, what we find historically is a process of primitive capital formation that developed over the course of centuries and snowballed into the capitalist production cycle, a phenomenon referred to as 'primitive accumulation' (Marx 1990; Perelman 2000; Federici 2005). 'Primitive accumulation', wrote Marx, 'plays approximately the same role in political economy as original sin does in theology', insofar as it was 'the historical process that separated the producer from the means of production' (Marx 1990, 873). Moore argues that the lines of primitive capital appropriation took the form of 'Cheap Natures' – free lunches for capitalism, in essence, in the form of cheap raw materials (extracted from third world countries), cheap labour (slaves owned and rented), cheap energy (the remains of dinosaurs converted by natural processes into a source of fuel for free) and cheap food (staples like bread and rice) (Moore 2015; 2016; 2017).

Where the Anthropocene approach suggests, as a tenet of nature understood in the abstract, that the Industrial Revolution was the beginning of a process that lead to the climate crisis, historical nature, on the other hand, understands the Industrial Revolution as the end of one – the end of the process of establishing and entrenching the social relations that would eventually create favourable conditions for the climate crisis in turn (Wallerstein 1986). Understood in this sense, we can understand the Industrial Revolution as the consequence of this process of primitive capital accumulation, gaining enough momentum to run under its own power; this historical context provides the basis for the integral consciousness of the oikeios. Marx described the opening of lines of free lunches for capital in the form of Cheap Natures to this end in the following terms:

> The discovery of gold and silver in America, the extirpation, enslavement and entomb-ment in mines of the aboriginal population, the beginning of the conquest and looting of the East Indies, the turning of Africa into a warren for the commercial hunting of black-skins, signaled the rosy dawn of the era of capitalist production. These idyllic proceedings are the chief moments of primitive accumulation. On their heels treads the commercial war of the European nations, with the globe for a theatre. . . . These methods depend in part on brute force, e.g., the colonial system. But, they all employ the power of the state, the concentrated and organized force of society, to hasten, hot-house fashion, the process of transformation of the feudal mode of production into the capitalist mode, and to shorten the transition. Force is the midwife of every old society pregnant with a new one. It is itself an economic power. (Marx 1990; 915–916)

The development of primitive accumulation as the original sin of the global capi-talist economy traces back at least as far as the fourteenth-century Europe, when a series of environmental and social catastrophes provided leverage opportunities for a powerful alliance of ecclesiastical and other privileged interests. European society was already in crisis at that time as the demands of supporting an idle manorial elite put ever-greater stress on production; goods that might have gone to supporting an expanding populace were diverted instead to maintaining an unproductive and gen-erally parasitical ruling class in the lifestyle to which they had become accustomed. These crisis conditions were exacerbated by the fact that peasants under feudalism were unable to raise productivity by innovating in the tools of production, having not the means to do so. Surpluses were expropriated anyway, an additional disincen-tive for the peasantry to produce any more than they needed for their own consump-tion (McNally 1990; Mielants 2008).

The first of these environmental and social catastrophes was a climate event known as the 'Little Ice Age', which gave rise to mass flooding throughout Europe that became known in turn, not entirely unsurprisingly, as the Great Flood (1314–1317). The flooding of the productive land resulted in widespread crop failure throughout Northern Europe; the inability of peasant farmers to dry the grain that could be harvested, resulting in turn in the exhaustion of the available stores. Crop failure and store exhaustion produced mass impoverishment, famine and the destructive effects of an inadequate diet on the physical resilience and immune system of the mass of the population – tinderbox conditions for a pandemic, which arrived with seeming inevitability in the form of the Black Death (1346–1353) (Cantor 2001, 9–10; Aberth 2013). As the Black Death killed sinner and believer alike, the European peasantry abandoned belief in a Divine Plan en masse in favour of class struggle and a contest over the future direction of the European economy (Hilton 1990; Aston & Philpin 1987; Wallerstein 2011).

The death of a third to a half of all of Europe had created a labour shortage which shifted the balance of power from landowner to labourer and enabled the steady breakdown of feudal bonds in the process, while the implications of what we might call the Epicurian Paradox provided incentives to revolt.

Is God willing to prevent evil, but not able? Then he is not omnipotent.
Is he able, but not willing? Then he is malevolent.
Is he both able and willing? Then whence cometh evil?
Is he neither able nor willing? Then why call him God?

Amid these class antagonisms, the Catholic hierarchy and other landed interests moved to defend their very earthly power against heterodoxy, apostasy and rebellion, born of loss of faith in God's Plan as the basic explanation for all life (and of church authority in interpreting it). Stripped of its enabling pretext, the class hierarchy that had tolerated the degeneration of environmental changes into famine and pandemic had become exposed (Hilton 2003; Cohn 2011). Thus in the name of defending their collective class interests, it had become necessary to construct a counter-narrative, a conspiracy theory alleging that the environmental disasters behind the Black Death were the work of witches, brides of satan, who poisoned the wells, destroyed crops and rendered men impotent. On this basis, they instigated the European Witch Hunts (approx. 1450–1750), spreading the witch conspiracy theory to shift the blame for the misery of the European peasantry under the feudal class structure onto its most vulnerable victims – poor peasant women who approximated the stereotype of the haggard old crone – and wage class war against dissent, heterodoxy and apostasy with a theocratic terror.

In revealing how the European Witch Hunts provided an enabling pretext for the terrorizing of the European peasantry – namely, liberating Europe from the malevolent influence of the witch – Federici also demonstrates how the Witch Hunts enabled crisis leveraging in the form of social engineering. Besides demonstrating to the peasantry what happened to those who failed to obey the theocratic class power, the Witch Terror expedited the reconstruction of the European class hierarchy by imposing patriarchal familial relations on peasant women, forcing them through overt threats to life and limb into the subordinate roles prepared for them within a resurgent regime of class rule – that is to say, as broodmares for capital. This, as it turns out, was a key form of primitive accumulation, and an essential facet of consolidating the foundations of the nascent capitalist economy (Federici 2005). The notable difference here was that the exploitation of cheap labour in the context of gender relations was carried out by means of appropriation rather than accumulation, dwarfing the value stolen through the exploitation of wage labour (Moore 2017a).

As it was developed by the dominant classes of the period between the end of the Late Middle Ages and the beginning of the Modern Era, the process of primitive accumulation took a variety of forms, including but not necessarily limited to the following, all of which involved either exploitation or appropriation:

1. Colonization of the feudal commons via enclosures, an act that forced the peasantry off the land, first into agrarian wage labour and then into the cities

to become wage slaves in industrial plants. (Marx 1990; Vol. I, Ch. 27; Boyle 2003; Perelman 2000; Thompson 1963; 1975)

2. Colonization of militarily conquered and resource-rich imperial possessions overseas for the exploitation of 'Cheap Natures' in the form of free land and the free labour of enslaved human resources. (Taussig 2010; Mies 2014; Moore 2015)

3. Colonization of the female body as a means of breeding factory fodder for exploitation in industry via the wage system and war fodder for the military acquisition of colonial possessions; the subjugation of women in general to a patriarchal social order in the name of raising an army of brood mares for capital and the state. (Federici 2005)

4. Colonization of individual subjectivity in the form of binary-laden ideological suppositions, that in replacing the individual personality structure with pre-made thoughts and ideas constituted in stereotypes and producing what social psychologists refer to as automaticity of behaviour, produced what we might regard as a form of Cheap Biopower. (Debord 1973; Bargh et al. 1996, 230; Blair & Banaji 1996, 1142; Moore 2015; Cisney & Morar 2015)

In all four examples, we find that binary logic accompanied the extension of primitive capitalist accumulation to all parts of the society until it reached a critical mass at the Industrial Revolution. The society versus nature binary enabled the dehumanization, subjugation and exploitation of workers – women workers in particular (whether paid a wage or not) and especially when not paid a wage as in the 'brood-mare for capital' role; Moore rightly points out this constitutes a form of appropriation that dwarfs wage exploitation (Moore 2015). It enabled the same towards the Oriental Other, as a building block of colonialism, and has ultimately enabled the objectification of oppressed peoples and classes, flora and fauna, and ultimately the planet itself, and their reduction to things valued only in terms of their exploitability for profit (Crenshaw 1991).

As the historical example of the European Witch Hunts reveals, the very brutal terror needed to break down female resistance to capitalist patriarchy, as one front in a general class war geared towards successful primitive accumulation, required binary-ridden pretexts as an institutional imperative. At the core of the witchcraft conspiracy theory was the belief that the sexuality of women was deviant enough to render them susceptible to seduction by satan ('nature'), a problem that needed to be corrected with the intervention of the righteous in the form of the witch trials ('society'). As an expression of the archetypal battle between good and evil, this takes on the society versus nature binary took full advantage of the cultural priming effects of the religious narrative around the battle between God and Satan for the souls of humanity.

If at the core of the theocratic terror of the European Witch Hunts was the binary logic dividing the world into believers versus heretics, such binaries were also key facets of the other fronts of the class war over the future socioeconomic development of the European society, all of which profoundly undermined the unifying oikeios in the name of the 'divide and conquer' strategy so vital to the maintenance of class-based hierarchies. Colonialism and slavery required the enabling binary of civilization versus barbarians, heathens and savages, cowboys and Indians, one that sought to explain away enclosures, colonial land theft and patriarchal terror as a moral failing on the part of the victims on the one hand, and as a service being done to them by their usurpers in bringing them the benefits of Christian civilization on the other (Said 1979; Davies et al. 1993). In the case of the enclosures, the binary logic of propertarianism lionized usurpers of the Commons as Lords, while maligning the dispossessed as criminals, beggars and thieves. In what Polanyi described as 'a revolution of the rich against the poor' and a 'uniform catastrophe', the victims were cast as somehow lacking in respect for the rights of others – the patent projection characteristic of this mentality an indication of the tenor of ideological rationalizations of private property to come (Boyle 2003).

In enabling this blame-shifting then, the society versus nature binary played a crucial role in expediting the rise of capitalist modalities to global dominance prior to the industrial revolution, predicated on the destruction of natural unity represented by the oikeios, at which point the primitive forms of accumulation and appropriation snowballed into private accumulation, industrial capitalism and imperialism. This development raised the privatization of benefits and socialization of costs to an organizing principle of society as such, as it did the propensity to, in turn, to reduce nature to an infinite resource and infinite garbage dump (Moore 2016). Carried over into the organizing principle of capitalist production, this reductionist belief in the possibility and desirability of endless growth as the basis of primitive accumulation also played a crucial role in expediting the conditions under which the climate crisis could develop.

The origin of the climate crisis in the origin of capitalist modalities suggests that responses to the former that also neglects to address or turn a blind eye to the latter are bound to fail in that they fail to meet Einstein's principle, as noted earlier, that meaningful responses to problems rise above the thinking that created them, restoring the natural unity of the oikeios in the process. The historical role of primitive accumulation in kick-starting capitalism, within which benefits are privatized, and costs are socialized while the cult of endless growth is prioritized above the capacity of the planet to sustain life, is well enough established. Liberal environmentalism, as expressed in the Anthropocene concept, fails to take account of this and the binary logic that facilitated land theft, genocide, large-scale terror, enslavement and numerous other crimes against humanity. As such liberal environmentalism cannot help but fail to articulate an ultimately meaningful response, since it has neglected to adequately conceptualize the parameters of the problem.

TOWARDS A RESPONSE

Acknowledging climate change without transcending the thinking that gave rise to it, as we have seen, appears to amount to offering answers to questions no one is asking; if this is true of the Anthropocene idea, it is most certainly true of market-based approaches to climate change, which in Germany have already given rise to 'neoliberals on bikes', as one of the co-founders of the Greens now describes them (Cyran 2011).

Other tendencies manage nevertheless to reproduce binary thinking in the name of the solution. As David McNally observes, the emergence 'out of the centuries-old competitive activities of merchants and manufacturers in rational pursuit of their individual economic self-interest' was the basis of the liberal view of the origins of capitalism.

> The rise of capitalism is thus explained in terms of the rise to prominence of the most productive, rational and progressive social groups – merchants and manufacturers. Not surprisingly, classical political economy came to be seen as an intellectual reflection of the ascendance of merchants and manufacturers and as a theoretical justification of their interests and activities. (McNally 1990, xi)

Paradoxically enough, this also appeared to be the interpretation of Marx and Engels in The Communist Manifesto, who waxed lyrical about the glorious doings of incipient capitalism.

> The means of production and of exchange, on whose foundation the bourgeoisie built itself up, were generated in feudal society. At a certain stage in the development of these means of production and of exchange . . . the feudal relations of property became no longer compatible with the already developed productive forces; they became so many fetters. They had to be burst asunder; they were burst asunder. (Marx & Engels 2002)

They were burst asunder with the aid of three centuries of state terror, as were the alternative paths of cooperative development springing up around parts of Europe where feudal bonds had ceased to have influence, mirroring as they did the Russian Obschina or Mir that in turn predated serfdom. Furthermore, according to this interpretation the bourgeoisie as the most progressive class did all the work, though as we know thanks to Marx and Engels' own later study, the work was done by chattel slavery, or under the waged variant, which in allowing the emerging capitalist class to free up capital costs associated with owning and maintaining the labour supply, which meant that slaves were no longer owned, but rented.

Mere facts such as these notwithstanding, the feeling that they had discovered underlying laws of historical development compelled Marx and Engels to invoke a binary between 'scientific' socialism and 'utopian' socialism, the difference according to Engels being that the 'scientific' variety was not

> an accidental discovery of this or that ingenious brain, but the necessary outcome of the struggle between two historically developed classes – the proletariat and the bourgeoisie. Its task was no longer to manufacture a system of society as perfect as possible, but to examine the historical-economic succession of events from which these classes and their antagonism had of necessity sprung, and to discover in the economic conditions thus created the means of ending the conflict. (Engels 1892)

According to this view, placing revolutionary theory and practice in historical context then meant coming to terms with this 'historical-economic succession of events' – one that nevertheless neglected to account for historical events producing capitalism, in fact, the European Witch Hunts being the prime example. If all history hitherto had been a history of class struggles, the outcome of the decline of feudalism had hardly been preordained, as historical materialism seemed to imply ('so many fetters burst asunder' etc.). The fact that this was implied indicated historical materialism was being deployed, not as a tool of disinterested understanding, but as a legitimizing ideological pretext for the capture of state power.

The cognitive dissonance within Leninism goes some way towards accounting for the fact that the tendency of the powerful to conflate criticism with attacks and 'existentialist threats'.[3] Similarly, it has also been the propensity of Stalin to conflate criticism with attacks on his rule from apologists for capitalist reaction. In The Results of the First Five-Year Plan (1933), Stalin utilized the False Dilemma ('those who are not for me are against me', or 'there is no difference between being criticized and being attacked') to neutralize dissent, alleging that

> we must bear in mind that the growth of the power of the Soviet state will increase the resistance of the last remnants of the dying classes. It is precisely because they are dying, and living their last days that they will pass from one form of attack to another, to sharper forms of attack, appealing to the backward strata of the population, and mobilizing them against the Soviet power. There is no foul lie or slander that these 'have-beens' would not use against the Soviet power and around which they would not try to mobilize the backward elements. This may give ground for the revival of the activities of the defeated groups of the old counter-revolutionary parties: the Socialist-Revolutionaries, the Mensheviks, the bourgeois nationalists in the centre and in the outlying regions; it may give grounds also for the revival of the activities of the fragments of counter-revolutionary opposition elements from among the Trotskyites and the Right deviationists. Of course, there is nothing terrible in this. But we must bear all this in mind if we want to put an end to these elements quickly and without great loss. (Stalin 1976)

Characteristic in this passage is the binary between 'the power of the Soviet state' and 'the last remnants of the dying classes', assuming a self-serving association of state power and the revolution, versus critics of the state power and reaction – one that would come to full fruition in the Stalinist Purges. Leon Trotsky adopted the same working assumption in alleging of critical tendencies within the early Bolshevik Revolution, including the Kronstadt sailors, that

> the Workers Opposition have come out with dangerous slogans. They have made a fetish of democratic principles. They have placed the workers' right to elect representatives above the Party. As if the Party were not entitled to assert its dictatorship even if that dictatorship temporarily clashed with the passing moods of the workers' democracy. (Mandel 1995, 83)

It was not Trotsky who had been corrupted by the exercise of absolute power; it was the fault of the Russian working class for asserting workers' democracy (Aufheben

2011, 6–46). Prior to this outburst, Trotsky had referred to them as the 'cream of the revolution' when their revolutionary tendencies served his purposes, before massacring them when their cream-like qualities became too much of an obstacle to his own ambitions.

Ironically enough, his own invocation of the False Dilemma involving legitimate revolutionaries and proponents of an existentialist threat to the revolution would be used against Trotsky when he became the leader of the Left Opposition to Stalin after 1924. This successful application of historical materialism, based on the kind of scientific understanding of socialism beyond the dilettantish revisionism of the utopian, was previously articulated by Lenin, who alleged that

> state capitalism, which is one of the principal aspects of the New Economic Policy, is, under Soviet power, a form of capitalism that is deliberately permitted and restricted by the working class. Our state capitalism differs essentially from the state capitalism in countries that have bourgeois governments in that the state with us is represented not by the bourgeoisie, but by the proletariat, who has succeeded in winning the full confidence of the peasantry. (Lenin 1922)

As Trotsky had already argued, citing the historical materialist conception of history, however, the Party was entitled to assert its dictatorship in opposition to the 'passing moods of the workers' democracy', even if doing so required the imposition of binaries that fatally undermined the oikeios. Understood in these terms, the New Economic Policy (NEP) could be accounted for 'in terms of the rise to prominence of the most productive, rational and progressive social groups – merchants and manufacturers'.

The difference in this case was the claim that words spoke louder than actions, and thus the justice of the NEP was a matter of what the Leninists introducing it claimed to believe, rather than what they did. Once again, conflating the exercise of state power with the mass of the people also meant conflating doubting of their judgement with attacks on the revolution. This fact raises the issue of the role of these underlying assumptions in helping to promote the coming climate crisis by promoting state capitalism, though some might perhaps protest that Soviet industrialization through the proletarian state also produced proletarian global warming since it, too, was a product of the Soviet power.

Applied to climate politics, the propensity of orthodox Marxists to conflate doubt in their judgement with attacks on social justice and the climate gives rise to outbursts such as that of John Bellamy Foster. In Foster's review by Jason W. Moore, which he happened not to like, he wrote, 'So I would not refer at all to "Moore's Marxism," except ironically,' adding that, 'the framework he has developed is anti-ecosocialist and anti-ecological'.

> I can only conclude that he has joined the long line of scholars who have set out to update or deepen Marxism in various ways, but have ended up by abandoning Marxism's revolutionary essence and adapting to capitalist ideologies. . . . No doubt Moore's work has attracted and will attract some notable scholars. But in terms of ecological

Marxism it is necessary to draw a line. Moore, I am sorry to say, has moved to the other side, and now stands opposed to the ecosocialist movement and socialism (even radicalism) as a whole. (Climate & Capitalism 2016)

Foster's superior airs notwithstanding, the irony of suggesting that someone who expresses a point of view that he happens not to like or agree with is given over to idealism and counter-revolutionary impulses is hard to miss. A further paradox is evident in the claim of 'Marxism's revolutionary essence', given the issues with Leninism referred to earlier, and Foster's association of ideas he likes with 'Marxism's revolutionary essence', along with ideas he doesn't like with active hostility to radicalism, socialism, and ecology. Foster's willing conflation of expressions of doubt in his judgement and attacks on his person is indicative of the same tendency in the Bolsheviks, as is their habit of identifying their own interests with those interests of the revolution, and of associating any threat to the one with a threat to the other. Such facts invite the conclusion that Foster, like his forebears, embodies the idealism, utopianism, and revisionism he attributes to his enemies, useful tools apparently in perpetrating cowardly and vicious attacks designed to compensate for a comparative lack of ideas and ideological coherence. Furthermore, it perpetuates the binary logic that undermines the oikeios, perpetrating the thinking that gave rise to global warming in the name of combating it. The same appears only truer of the binary logic evident at the origins of the climate crisis in historical nature; the insights that reveal the limitations of Crutzen's Anthropocene idea and Neoliberals on Bikes thus also reveal those of Leninism and its ecological offshoots. The limitations for each are similar –they each fail to transcend the thinking that created the problem and, in so doing, reproduce in practice what they purport theoretically to oppose.[4] Even the motto of Climate & Capitalism, the website on which Foster's attacks appear, is 'eco-socialism or barbarism: there is no third way' – 'if you doubt the judgment of an eco-socialist', in other words, 'the barbarians win'.

FOR WORKERS' CONTROL

As we have seen, the European economy spawned capitalism on the ruins of feudalism as the privileged classes successfully fought to reassert their class dominance in the face of attempts by the productive classes to control the conditions of their own lives. This problem is one that the Bolsheviks also faced in their campaign to assert theirs in the name of socialist revolution. In calling attention to the contended nature of the European economy at the end of the feudal era, Federici inadvertently exposes Leninist apologetics for state capitalism by revealing the pseudo-scientific, deterministic foundation of historical materialism. That the Catholic hierarchy did take the reconstruction of class power seriously enough to wage 300 years of theocratic wars in the name of suppressing threats to it highlights the concerted and protracted social engineering behind the imposition of capitalist social relations – a campaign that was anything but spontaneous.

In their ignorance, willing or otherwise, of the actual basis for the development of capitalist social relations, the Bolsheviks became subject to the capitalist modalities that give rise to them in the form of the binary thinking that creates the possibility of objectification and exploitation – a fact that goes some way towards accounting for Lenin's tawdry apologetics for state capitalism. Insofar as liberal and authoritarian socialist responses to the climate crisis share the problem of reproducing the capitalist modalities that give rise to them in the form of the binary thinking that creates the possibility of objectification and exploitation, they also share the problem of a basic inconsistency between means and outcomes. The ultimate expression of this inconsistency is the exercise of state power, which follows its own logic regardless of who possesses it. As James Madison argued, 'The primary function of government is to protect the minority of the opulent from the majority'; to this, we might add, the secondary function of government is to ensure that those who are now in power stay in power. Bolsheviks and Neoliberals on Bikes alike heartily agree on the necessity of the state; as the history of these loyal opponents demonstrates, both produce outcomes consistent with the values they apply, not the ones they profess.

The impetus to produce outcomes consistent with values professed as well as those applied, on the other hand, has produced the spectrum of strategies and approaches from the libertarian socialist corner of the political compass – one inclusive of schools as varied as anarcho-communism, platformism, municipalism, anarcho-syndicalism, autonomism, council communism, and libertarian Marxism (Biehl & Bookchin 1998; Pannekeok 2003; Rocker 2004; Solidarity Federation 2012; Mattick 2017). The strategic and moral imperative to maintain a harmony between means and outcomes (and thus also words and actions) dovetails with the argument that environmentally sustainable production stands the greatest chance of succeeding when carried out under direct community control for the public, rather than under private control for private gain; it anticipates that that production will be carried out rationally by virtue of being under the self-management of those who perform it, who is having to live with the consequences of their choices on that count will thus be compelled to bear responsibility for them.

As one of the more theoretically and strategically robust schools of libertarian socialism, anarcho-syndicalism looks to maintain a harmony between means and outcomes by proposing to shift workers' struggle from the political sphere, where it is weakest, to the point of production, where it is strongest. Revolutionary industrial unions and confederations of industrial federations, coordinated using mandated recallable delegates, would serve as a 'practical school of socialism'. Here, day-to-day struggles to defend rights and advance interests would act as a form of 'revolutionary gymnastics', in preparation for the day when the opportunity arose to take control of the means of production and establish workers' control, such as in the case of a revolutionary general strike. As Rudolf Rocker argued,

> Only in the realm of the economy are the workers able to display their full spirit, for it is their activity as producers which holds together the whole social structure, and guarantees the existence of society at all . . . Education for socialism does not mean for them

trivial campaign propaganda and so-called 'politics of the day', but the effort to make clear to the workers the intrinsic connections among social problems and, by technical instruction and the development of their administrative capacities, to prepare them for their role of reshapers of economic life, and give to them the moral assurance required for the performance of their task. (Rocker 2004, 58)

Before many of them were destroyed by the Stalinists fulfiling the historical destiny of scientific socialism, these ideas gave rise to the agrarian and industrial collectives created during the Spanish Revolution, which commenced on 19 July 1936. The attempted coup by the eventual victor in the civil war, Francisco Franco, created a political and social vacuum filled by the creative organizational spontaneity of Spanish workers, who collectivized industries throughout the Republican areas and began running them without managers for need rather than profit. In Catalan metalworking firms, for example, and as one CNT-FAI bulletin announced,

> As a result of the events of July, two new forms of administration have surfaced. One, involving worker management without any restrictions whatsoever, by means of takeover. The other represents a greatly attenuated bourgeois mode of administration through monitoring activity carried out by workers' factory committees. (Mintz 2013, 66)

In Barcelona, to take another, transportation services were collectivized. Administration and timetabling were rationalized, wages were standardized, hours were reduced to provide additional employment for those out of work, and pensions were arranged for retirees (Mintz 2013, 68–69). In the countryside of Catalonia, the Levante and Aragon, some 3 million peasants on the land collectivized agricultural production, a fact for which scientific socialism is yet to account (Leval 1975; Ness & Azzellini 2011). Whether in industry or on the land, the new mode of production adopted forms consistent with the outcomes desired, allowing for further progress towards full collectivization where it had not yet already been achieved, while providing industrially democratic mechanisms for all involved to exercise a meaningful measure of control over the conditions of their own work, and to have input into how and for what purpose production was carried out.

> The structure of the new economy was simple. Each factory organized a new administration manned by its own technical and administrative workers. Factories in the same industry in each locality organized themselves into a Local Federation of their particular industry. The total of all the Local Federations organized themselves into the Local Economic Council in which all the centres of production and services were represented: coordination, exchange, sanitation and health, culture, transportation, etc. Both the local federations of each industry and the Local Economic Councils were organized regionally and nationally into parallel National Federations of Industry and National Economic Federations. (Dolgoff 1974, 66)

Ness and Azzellini (2011) rightly locate the collectivizations carried out during the Spanish Revolution as one of a series of achievements realized throughout the twentieth century. Citing alongside this achievement, among others, Ness and Azzellini (2011) include the industrial democracy of the factory committees in the early days of the Russian Revolution, before the consolidation of the Bolshevik state, the Italian factory occupations of 1920, and the forms of workers' self-management established in the former Yugoslavia. By providing for the individual freedom of the worker and their ability to control the conditions of their own work, such episodes provided for the emancipation of the workers from the oppression of class privilege in practice, as well as in rhetoric. Insofar as they did this,[5] they might be considered responses to such problems that, in terms of their economic achievements at least, successfully transcended the thinking that created them.

CONCLUSION

In failing to examine the historical origins of capitalism, as we have seen, liberal responses to climate change neglect to account for the corporate supremacist nature of neoliberal ideology, as well as the forces and tendencies that have given rise to it historically, and the assumption that the world is an infinite resource and infinite garbage can as a characteristic feature (as opposed to something that can be reformed away). In so doing, they cede history to those responsible for the problem – limiting any possibility of using history to surmount it.

Similarly, scientific socialism, neglecting to account for the alliance of privileged forces that brought global capitalism into being and the climate crisis along with it at the end of the Feudal era, has also failed to overcome the Faustian Bargain involved in resorting to binary logic as crisis leveraging. In the case of early capitalists, as we saw, this meant demonizing those they sought to exploit and usurp in pursuit of primitive accumulation, and anyone who got in their way. In this respect, the Bolshevik witch-hunting of their opponents via the bloody repression of the Kronstadt Rebellion and the Great Purges reveals that while they might have believed they could have their cake and eat it too, in the end, the Faustian Bargain they made with the state exacted their soul as its price.

At issue here ultimately was the fact that, as Eugene Debs pointed out, those who lead the workers into a revolution can lead them back out again. Not only is the emancipation of the working class from the oppression of serving-class privilege desirable, democratizing the economy and bringing it under the direct and collective control of producers is critical from the point of view of making the economy sustainable, if we consider that that difference between an economy driven by profit and one driven by need is the difference between an economy where those who are responsible have to live with the consequences of their choices the one we have at the moment, where they do not (Klein 2016). The capacity of purportedly liberal states for witch-hunting fared no better (Feldman 2011). As the archetype for later Show Trials and Purges, whether in Moscow or Hollywood, the European Witch Hunts also constitute a precedent; in addition to revealing much about the actually existing operating principles of the state as an institution of class privilege, the ubiquity of their binary logic reflects the consistency of purpose in undermining knowledge and understanding of natural unity embodied in the oikeios concept.

It is in the resurgent oikeios, however, that the climate crisis can now be understood as a facet of the general crisis of civilization, interlinked with economic crisis (wealth distribution), social crisis (austerity) and political crisis (corporate capture). The problem of overcoming the thinking that gives rise to the crisis of civilization becomes a matter of transcending the binary thinking of autocratic, hierarchical idealism and reasserting the intersectional logic of the oikeios. At this late hour of late capitalism, the possibility of reestablishing mass workers' organizations, including revolutionary unions that can proactively declare a revolutionary general strike and take control of the means of production, seems remote; we might better rely on a strategy of 'reconstruction', of catching society as the dominant institutions teeter and keel over and resuming production on the basis of workers control – as has been done in a number of notable cases already (Magnani 2003; Klein & Lewis 2004). In this sense, 'reconstruction' has a literal meaning of reconstructing the physical fabric of society but also a metaphorical sense of reconstruction of the intellect out of the rot and decay of neoliberal ideology.

Either way, if the desire to avoid reproducing the thinking that created the problem of climate change factors into our thinking at all, then establishing and maintaining a harmony between means and outcomes must be paramount. Workers' control, as the crucial basis for sustainable production, must be reflected in the values we apply in fighting for it, not just in those we profess. The imperative to take seriously the issues surrounding the role of binary thinking in the creation of the problem, and its incorporation into responses that reproduce what they claim to oppose, was expressed originally and best in the motto of the First International that 'the emancipation of the working class shall be carried out by the workers themselves' – this is to say, by the workers directly, not by those claiming to speak in their name. This distinction is no longer simply a question of justice, or even of protecting the movement from opportunists and usurpers; it is now also one of survival.

NOTES

1. This chapter contends that the constructive achievements in workers' self-management of production realized during the Spanish Revolution and Civil War (1936–1939) constitute a prime example.

2. The Political Compass rightly places political philosophies on two axes ('libertarian vs. authoritarian' and 'collectivist vs. individualist'). For more see politicalcompass.org.

3. Which is also a feature of Whitch Hunts.

4. One can only wonder at the contortions of logic that would have been necessary had the Russian Revolution taken place 100 years later than it did.

5. As well as pointing at the same moment towards a potential way out of the problems created by an economic logic devoted to infinite growth based on the privatization of benefits and socialization of costs.

REFERENCES

Aberth, J. (2013). *From the Brink of the Apocalypse: Confronting Famine, War, Plague and Death in the Later Middle Ages*. London: Routledge.

Angus, I. (2016). *Facing the Anthropocene: Fossil Capitalism and the Crisis of the Earth System*. New York: New York University Press.

Aston, T. H., & Philpin, C. H. E. (Eds.). (1987). *The Brenner Debate: Agrarian Class Structure and Economic Development in Pre-Industrial Europe*. Cambridge, MA: Cambridge University Press.

Aufheben Collective. (2011). *What Was the USSR? Towards a Theory of the Deformation of Value Under State Capitalism*. Edmonton: Thoughtcrime Ink.

Bargh, J. A., Chen, M., & Burrows, L. (1996). Automaticity of Social Behavior: Direct Effects of Trait Construct and Stereotype Activation on Action. *Journal of Personality and Social Psychology* 71, no. 2: 230–244.

Biehl, J., & Bookchin, M. (1998). *The Politics of Social Ecology: Libertarian Municipalism*. Montreal: Black Rose Books.

Blair, I. V., & Banaji, M. R. (1996). Automatic and Controlled Processes in Stereotype Priming. *Journal of Personality and Social Psychology* 70, no. 6: 1142.

Boyle, J. (2003). The Second Enclosure Movement and the Construction of the Public Domain. *Law and Contemporary Problems* 66, no. 1/2: 33–74.

Brinton, M. (1970). *The Bolsheviks & Workers' Control*. London: Solidarity.

Cantor, N. F. (2001). *In the Wake of the Plague: The Black Death and the World It Made*. New York: Simon and Schuster.

Cisney, V. W., & Morar, N. (Eds.). (2015). *Biopower: Foucault and Beyond*. Chicago: University of Chicago Press.

Climate & Capitalism. (2016). *In Defense of Ecological Marxism: John Bellamy Foster Responds to a Critic*. http://climateandcapitalism.com/2016/06/06/in-defense-of-ecological-marxism-john-bellamy-foster-responds-to-a-critic.

Cohn, N. (2011). *The Pursuit of the Millennium: Revolutionary Millenarians and Mystical Anarchists of the Middle Ages*. London: Random House.

Cox, C. R. (2015). Faulty Presuppositions and False Dichotomies: The Problematic Nature of 'the Anthropocene'. *Telos*, no. 172: 59–81.

Crenshaw, K. (1991). Mapping the Margins: Intersectionality, Identity Politics, and Violence Against Women of Color. *Stanford Law Review*, 1241–1299.

Crutzen, P. J. & Stoermer, E. F. (2000). The 'Anthropocene'. *Global Change Newsletter* 41.

Cyran, O. (2011). Neoliberals on Bikes. *Counterpunch*. https://www.counterpunch.org/2011/09/30/neoliberals-on-bikes.

Davies, M. W., Nandy, A., & Sardar, Z. (1993). *Barbaric Others: A Manifesto on Western Racism*. London: Pluto Press.

Debney, B. (2017a). Crises Worthy and Otherwise: Terror Scare vs Climate Change. *Counterpunch* 24, no. 2.

———. (2017b). Historical Nature Versus Nature in General: Capitalism in the Web of Life. *Capitalism Nature Socialism* 28, no. 2: 126–131.

Debord, G. (1973). *The Society of the Spectacle*. Detroit: Black and Red.

Deckard, S. (2009). *Exploiting Eden: Paradise Discourse, Imperialism, and Globalization*. London: Routledge.

Dolgoff, S. (Ed.). (1974). *The Anarchist Collectives: Workers' Self-Management in the Spanish Revolution, 1936–1939*. Montreal: Black Rose Books.

Engels, F. (1892). *Socialism, Utopian and Scientific*. New York: Labor News Company.

Federici, S. (2008). *Caliban and the Witch: Women, the Body and Primitive Accumulation*. New York: Autonomedia.

Feldman, J. (2011). *Manufacturing Hysteria: A History of Scapegoating, Surveillance, and Secrecy in Modern America*. New York: Pantheon.

Hilton, R. (Ed.). (1998). *Class Conflict and the Crisis of Feudalism: Essays in Medieval Social History*. London: Verso.

Hilton, R. (2003). *Bond Men Made Free: Medieval Peasant Movements and the English Rising of 1381*. London: Routledge.

Intergovernmental Panel on Climate Change (IPCC). (2008). Fifth Assessment Report (AR5). https://www.ipcc.ch/report/ar5/.

Klein, N. (2016). Let Them Drown: The Violence of Othering in a Warming World. *London Review of Books* 38, no. 11·(2 June): 11–14.

Klein, N., & Lewis, A. (2004). The Take: Occupy! Resist! Produce! First Run Features:Icarus Films.

Lenin, V. I. (1987). To the Russian colony in North America. In *Completed Works of Lenin* (Vol. 43, pp. 290–291). Beijing: People's Publishing House.

Leval, G. (1975). *Collectives in the Spanish Revolution*. London: Freedom Press.

McNally, D. (1990). *Political Economy and the Rise of Capitalism: A Reinterpretation*. Berkeley, CA: University of California Press.

Magnani, E. (2003). *El Cambio Silencioso: Empresas y fábricas recuperadas por los trabajadores en la Argentina*. Buenos Aires: Prometeo libros.

Mandel, E. (1995). *Trotsky as Alternative*. London: Verso.

Marx, K. (1990). *Capital: Vol. 1*, Ben Fowkes (trans). London: Penguin Classics.

Marx, K., & Engels, F. (2002). *The Communist Manifesto*. London: Penguin Classics.

Mattick, P. (2017). *Anti-Bolshevik Communism*. London: Routledge.

Mielants, E. (2008). *The Origins of Capitalism and the "Rise of the West."* Philadelphia, PA: Temple University Press.

Mies, M. (2014). *Patriarchy and Accumulation on a World Scale: Women in the International Division of Labour*. London: Zed Books.

Mintz, F. (2013). *Anarchism and Workers' Self-Management in Revolutionary Spain*. Oakland: AK Press.

Moore, J. W. (2015). *Capitalism in the Web of Life: Ecology and the Accumulation of Capital*. London: Verso.

Moore, J. W. (Ed.). (2016). *Anthropocene or Capitalocene? Nature, History, and the Crisis of Capitalism*. Oakland: PM Press.

Moore, J. W. (2017). The Capitalocene, Part II: Accumulation by Appropriation and the Centrality of Unpaid Work/Energy. *The Journal of Peasant Studies*: 1–43.

Moore, J. W., & Patel, R. (2017). *A History of the World in Seven Cheap Things a Guide to Capitalism, Nature, and the Future of the Planet*. Berkeley, CA: University of California Press.

Ness, I., & Azzellini, D. (2011). *Ours to Master and to Own: Workers' Control from the Commune to the Present*. Chicago: Haymarket Books.

Pannekoek, A. (2003). *Workers' Councils*. Edinburgh: AK Press.

Perelman, M. (2000). *The Invention of Capitalism: Classical Political Economy and the Secret History of Primitive Accumulation*. Durham, NC: Duke University Press.

Rocker, R. (2004). *Anarcho-Syndicalism: Theory and Practice*. Oakland: AK Press.

Runehov, A., Lluis Oviedo, L. C., & Azari, N. P. (Eds.). (2013). *Encyclopedia of Sciences and Religions*. New York: Springer.

Saïd, E. (1979). *Orientalism*. New York: Vintage.

Solidarity Federation. (2012). *Fighting for Ourselves: Anarcho-Syndicalism and the Class Struggle*. London: Freedom Press.

Stalin, J. V. (1976). *Problems of Leninism*. Moscow: Foreign Languages Press.

Taussig, M. T. (2010). *The Devil and Commodity Fetishism in South America*. Chapel Hill: University of North Carolina Press.

Thompson, E. P. (1963). *The Making of the English Working Class*. London: Vintage.

Thompson, E. P. (1975). *Whigs and Hunters: The Origin of the Black Act*. London: Allen Lane.

Wallerstein, I. (1986). The Industrial Revolution: Cui Bono? *Thesis Eleven* 13, no. 1: 67–76.

Wallerstein, I. (2011). *The Modern World-System I: Capitalist Agriculture and the Origins of the European World-Economy in the Sixteenth Century*. Berkeley, CA: University of California Press.

Williams, G. A. (1975). *Proletarian Order*. London: Pluto Press.

Williams, R. A. (2012). *Savage Anxieties: The Invention of Western Civilization*. London: Macmillan.

10

The Soft Hand of Capital

Deric Shannon and Clara Perez-Medina

Picture if you will, what could be a regular day in a developed nation somewhere on Earth. Let's call it Sustainlandia. People wake up in the morning and are shuffled off on solar-powered light rails or perhaps electric, self-driving cars to their various jobs. They land in their participatory workplace and debate with workmates the workload of the day and how it will be divided. Once a consensus has been reached, they dole out tasks, taking care that labour is divided fairly. Sitting in plush workspaces, perhaps on hemp bean bags or with ergonomically designed desks, they work through their day with ample breaks for mindfulness training and meditation on the green roof installed on the office building. At the end of the workday, they return via light rail or electric car to their living units. These green buildings are designed to capture carbon, insulated by being burrowed partially in the Earth, and the spaces run on 100 per cent renewable energies. For those who don't earn quite enough from their jobs, there is a universal basic income that amounts to basic upkeep. Health care, education, fresh fruits and vegetables, and other socially agreed-upon goods have been declared rights, guaranteed by the state, and the income people receive from work, and the universal basic income is used for personal edification – life's necessities are provided for people with a little bit left over for living.

Far away, or perhaps in the same country, more likely somewhere in the Global South, a permanent underclass struggles to provide the necessities for the mindful and participatory lives of their sustainable counterparts. The risks we have come to associate with industrial life have skillfully been moved, out of sight and concern of those living in Sustainlandia. The collection of plastics, styrofoam and other wastes from a bygone era that was not shot into space is entered into landfills far away from the managerial population centres of Sustainlandia. People here, let's call it Ironborough, live off the scraps that emerge from Sustainlandia, providing for it fundamental goods. Someone, after all, must mine for the materials necessary for solar panels, wind farms and the digital machinery that allows for the production of services and goods in Sustainlandia. Those who manage to find non-automated work labour for long hours securing the needs of their managerial peers with intermittent access to life's necessities – to say nothing of the things one might desire to lead a *decent* life. This backdrop is hidden from view. The citizens of Sustainlandia need not think about life in Ironborough. There is time, space – epistemic distance – allowing the managerial elite to go about their mindful, green, sustainable lives without considering the costs foisted off onto parts of the world they never need visit.

In this piece, we argue that radicals have recognized what is taking place in Ironborough, perhaps to the exclusion of what might become Sustainlandia. That is, we have typically looked critically – and rightfully so – at capitalism's iron fist, but too much to the exclusion of its soft hand. Because of this, we have missed out on the co-opting possibilities in the discourses of (and practices related to) sustainability. Through highlighting the possibilities of sustainable capitalism, we argue that anticapitalists should re-engage with both political economy and environmentalism with a political ecology rooted in anarchism's critique of relations of domination. That is, capitalism has any number of possible outcomes. Radicals have typically argued in political-economic and environmental tonnes that embrace a sort of millenarianism. This necessarily leads to the belief that capitalism must be opposed because it will destroy itself (either through class struggle, war or environmental decline). We maintain that Sustainlandia and its counterpart, Ironborough, are also possibilities for the development of global capitalism. That perhaps capitalism is *not* on a collision course with itself and that it might be managed and maintained to the benefit of a class of global managers at the expense of a crushed and compliant working class. Perhaps green capitalism (for some) is possible.

We begin by outlining some well-known examples of political-economic and environmental millenarianism, both historical and contemporary. Next, we offer some large-scale examples of attempts at creating a sustainable capitalism and invite readers to think about what it might mean to scale these experiments up globally. In doing so, we point out that there is some possibility that capitalism can create its own release valves, political-economic and environmental safety nets that rescue capitalism from its own worst tendencies – it can do so especially effectively through the power of the state. We should be aware, and wary, of the possibilities for the soft hand of capital to produce a green version of itself, effectively co-opting responses to its iron fist. Finally, we argue that if it is indeed one possibility for capitalism to develop a sustainable version of itself, radical political ecology should be rooted in an ethical critique of capitalism (that it is a miserable system) rather than an eschatological belief that we must oppose it or face the end of the world.

Our point is not to under-emphasize serious existential questions about the capacity of our current social-political-economic systems to continue on in the same fashion. Indeed, it seems with popular writings on the environment these days, it is standard to start with the familiar laundry list of environmental abuses and disasters: that is, ice sheets disappearing at increasing rates; a sixth mass extinction event already well underway; dangerous levels of plastic in the ocean; and a record-breaking 2017 hurricane season that ravaged many parts of the world. Of course, today, the circumstances seem even more direr: megalomaniacal rulers promising nuclear war, enacting racial terrorism and standing by as climate change indicators begin to cross thresholds from which we cannot retreat. We are not positing the likelihood of a capitalist and statist transition to greener versions of themselves, but rather intend to evaluate the assumptions that catastrophist claims are premised upon. But even things that are not likely might be possible, and the problem of declining ecological systems and increasing social stratification allow a chance for radicals to re-examine our foundational assumptions. For us, it is not difficult to imagine a more diverse, kinder and less apologetic state apparatus leading the next capitalist transition to Sustainlandia, with Ironborough toiling on in the margins to support it.

HISTORICAL MILLENARIAN THEORISTS: MARX AND LUXEMBOURG

One common eschatological theory put forward by radicals of various sorts can be found in Karl Marx's critique of political economy. A simplified version of Marx's view is that capital, in its rapacious need for profit and growth, has set the world's workers, the proletariat, on a collision course with the owners of means of production, the bourgeoisie and their representatives in the state apparatus. As Marx penned with his companion, Engels, '[s]ociety as a whole is more and more splitting up into two great hostile camps, into two great classes directly facing each other – Bourgeoisie and Proletariat' (Marx & Engels, 1848).

According to the theory, class struggle is the engine of historical progression, and it is less an ethical necessity for workers to struggle against their position as social subordinates so much as it is a *scientific* reality that Marx was merely observing, watching historical epochs turn through processes of class struggle. In fact, the 'history of all hitherto existing societies is the history of class struggles' (Marx & Engels, 1848). Capitalism, for Marx, was a progressive force in this developmental process, as historical epochs churned through the roiling process of classes in struggle for dominance.

These historical periods resulted from changes in the mode of production, as class struggle produced different productive capacities and possibilities. But while capitalism was a progressive force, moving societies from feudal forms to bourgeois forms, it contained within it the seeds of its own destruction. As class struggle was simplified under the rule of capital to the battle between the proletarians and the bourgeoisie, a reckoning was on the horizon. Indeed, in the process of class struggle, it was in the material interests of the owners of society to extract every bit of surplus value that they could from their workers and capital investments. So while the means of production under capitalism created the capacity to produce and distribute for human need instead of profit, workers were paid less and less to line the pockets of the capitalists.

The contradiction between labour and capital was starkest when considering this productive capacity reached under capitalism. Again, this stage in human development had created the possibility of producing enough to meet human needs. But this possibility needed to be measured against the continued immiseration of workers. Those who owned the means of production need not use them, but those who used the means of production did not own them. A tiny class of owners were enriched with a state apparatus managing their interests while the world's workers must sell their labour in order to have access to life's necessities. These internal contradictions fuelled the class struggle, which would result in the institution of a new mode of production with productive relations planned for human need rather than the profits of a few – communism.

This mechanical reading of Marxism is perhaps most evident among contemporary radicals who argue for a version of what they call 'accelerationism' (see, e.g., Mackay & Avanessian, 2014), though the uses of this term seem to be under debate.[1] And it might be best encapsulated with Marx's theory of the tendency for the rate of profit to fall (Marx, 1981: chapter 13, passim). According to this theory, the capacity for profit has necessary limits. A primary principle of political economy under this view is that the rate of profit continually falls, and capitalists will eventually no longer be able to produce profit as a result. This, of course, would lead to the end of capitalism, a system rooted in profit and growth, were it true.

While Marx's critique of capitalism, and his concomitant theory of its demise, was focused centrally on political economy, he was also attentive to capitalism's tendency towards environmental degradation. Marx (1981: 949) wrote of the possibility of capitalism to create an 'irreparable rift in the interdependent process of social metabolism'. He wrote of 'social metabolism' to point out that humanity must go through processes of mixing our labour with the natural environment, creating social relationships between and among humans *as well as nature*.

Bellamy Foster (2000), who coined the term 'metabolic rift' in reference to the problems that arise out of these social relations, argues that Marx's *social metabolism* can, in part, be applied to soil degradation. The split between town and country and the need for greater fecundity in the soil in order to produce for capital accumulation incentivized soil depletion in search of immediate gains. It also removed masses of people from being able to contribute to soil fecundity by densely populating cities (where human waste was thus concentrated), removing people (and their waste) from agricultural centres. These dual processes demonstrated the tendency for capitalist agriculture to lead to (in this case) soil depletion.

Marx noted tendencies within capital towards its own destruction. This was true in political-economic terms, where Marx believed capitalism created its own gravediggers in proletarians (who are yet to dig its grave 150 years later, but we digress). Capitalism, because it is a system based on capital accumulation and profit, must have limits because the rate of profit, according to Marx, had a tendency to fall. He also noted capitalism's capacity to alienate people from nature, creating a rift in the social metabolism and degrading our natural world. Capitalism, in short, was on a collision course with itself. Perhaps no Marxist put it in as famously stark terms as Rosa Luxembourg.

Luxembourg: Socialism or Barbarism

Luxembourg was an active revolutionary in the Social Democratic Party (SPD) of Germany in the early 1900s. In the lead-up to the First World War, anti-capitalists around the world demonstrated against imperialist war, arguing that workers must unite and fight against their common enemy in the bourgeoisie rather than turning towards nationalism and, eventually, against *each other* in capitalist wars. This global anti-war movement was to become a shell of itself as people answered the siren song of national identity and chauvinism. The opposition to the First World War was a failure, and even 'the SPD [the "socialist" Party Luxembourg belonged to] deputies in the German Parliament, the Reichstag, voted for the war credits to fund the Kaiser's armies' in August of 1914 (Campbell, 2014).

This historical experience took its toll on many early twentieth-century anti-capitalists. Indeed, even in the anarchist milieu – notorious for its anti-statism – the famous 'gentle anarchist', Peter Kropotkin (among others) betrayed his anti-nationalist principles to argue for an allied victory in the war in the infamous 'Manifesto of the Sixteen' (Kropotkin & Grave, 1916). This development in the international workers movement devastated Luxembourg. As the World War unfolded around her, workers killed their fellow workers in battles between nations instead of the Great War between *classes*.

It makes sense to quote Luxembourg at length here. To her credit, it is difficult to imagine a materialist analysis that was not, at the least, pessimistic in the mid of the horrors of a global war. And she does not concede to that pessimism, but rather sees the future unfolding in a binary pattern – either humanity will liberate itself, or it will destroy itself.

Friedrich Engels once said, 'Bourgeois society stands at the crossroads, either transition to socialism or regression into barbarism.' What does 'regression into barbarism' mean to our lofty European civilization? Until now, we have all probably read and repeated these words thoughtlessly, without suspecting their fearsome seriousness. A look around us at this moment shows what the regression of bourgeois society into barbarism means. This world war is a regression into barbarism. The triumph of imperialism leads to the annihilation of civilization. At first, this happens sporadically for the duration of modern war, but then when the period of unlimited wars begins, it progresses towards its inevitable consequences. Today, we face the choice exactly as Friedrich Engels foresaw it a generation ago: either the triumph of imperialism and the collapse of all civilization as in ancient Rome, depopulation, desolation, degeneration – a great cemetery – or the victory of socialism, which means the conscious and active struggle of the international proletariat against imperialism and its method of war. This is a dilemma of world history, an either/or; the scales are wavering before the decision of the class-conscious proletariat. The future

of civilization and humanity depends on whether or not the proletariat resolves man-
fully to throw its revolutionary broadsword into the scales. In this war, imperialism
has won. Its bloody sword of genocide has brutally tilted the scale towards the abyss
of misery. The only compensation for all the misery and all the shame will be if
we learn from the war how the proletariat can seize mastery of its own destiny and
escape the role of the lackey to the ruling classes (Luxemburg, 1915: Chapter 1).

Clearly, in the throws of the historical defeat of the proletarian movement in Germany and the advent of the the First World War, Luxembourg, like her predecessors,
saw only two possibilities for the future – the collapse of capitalism (and with it
civilization) on the one hand, and socialism on the other. Capitalism was careening
towards catastrophe.

CONTEMPORARY RADICAL MILLENARIANS:
KLEIN AND BOOKCHIN

Similar catastrophist claims loom in the contemporary public sphere. Political-
economic writer Naomi Klein popularized what Sandler (1994) calls the 'grow or
die' (GOD) theoretical position in her book *This Changes Everything: Capitalism vs.
the Climate*, positing that the current configuration of capitalism must inevitably
destroy the life processes necessary for social reproduction. Here she brings capitalism (although a mangled and incoherent version of it) into the fore of public discussions about climate change and ecological destruction. Throughout her book, she,
usefully, spells out the often-obscured connections between global political economy
and the environment and lays bare the crisis we find ourselves in. Klein constructs a
portrait of a rogue capitalism hell-bent on destroying the planet, detailing conservative media manipulation around climate change to the effects of global trade agreements on environmental and social health.

Klein shows her hand throughout the work, arguing ultimately that our collective
ecological cliff is a consequence of our reigning economic paradigm, 'deregulated
capitalism' (75), and calls for action outside of the 'rules of capitalism as they are
currently constructed' (88). Klein's foe is capitalism wrought from the neoliberal
age: one of privatization, deregulation and harsh austerity for the working class. She
unsurprisingly posits a more tightly regulated capitalism in which corporations are
held accountable to their nation's taxes, military budgets are slashed, and carbon
is properly taxed in order to fund the generous social safety nets necessary for an
almost-certain future of heightened eco-disasters (perhaps a bit like Sustainlandia).
In order to deal with the twin crises of the environment and political economy, she
writes, we must 'change everything'. By everything, it is clear she means the state's
particular relation to the contemporary capitalist ecological regime – falling short of
indicting capitalist social relations per se.

Despite her defanged anti-capitalist politics, she posits two ultimate situations that
are worth contending with: an unregulated capitalism that inevitably undermines its
ecological basis and life as we know it, and a Marshall Plan for the Earth that brings

about the massive levels of investment needed to forge a greener capitalism, and a softer and perhaps more visible state hand (although she's unlikely to admit as much). We think she's right about two things: one, that an unregulated capitalism could bring the biosphere and the foundations of production to a critical point from which they may not recover, and two, that a nicer, more socially conscious capitalism could be just the thing that 'changes everything' if by that we mean 'creates a softer and kinder exploitative capitalism'.

Klein understands the power of the state to curb capitalism's excesses and bring its systems in balance with the crises of its time (that are, oftentimes, created by its very political-economic conditions). Time after time Klein cites the massive infrastructural investments and shifts necessary (and possible) to avert a crisis before crossing the hard planetary deadlines we know we find ourselves pressed up against (and in some cases have already crossed, like the beginning of the collapse of the Antarctic ice shelf). 'Climate action', writes Klein, 'is, in fact, a massive job creator' and could 'generat[e] plentiful, dignified work, and radically rei[n] in corporate greed', – and we don't think she's off-target (124–25). When Klein proposes that *alterations* to our current political economy are on the horizon and not outside of the realm of capitalist social relations, we think she's right. Indeed, that is a constituent part of our thesis – another capitalism is possible (putting aside, for now, the question of its *probability*).

Bookchin's Grow or Die Position

Perhaps the most salient contemporary example of the GOD thesis among anarchists and radical ecologists was put forward by the social ecologist Murray Bookchin. Bookchin began his political life as a party Marxist before eventually declaring himself an anarchist. Later in his life, he renounced anarchism but nevertheless persisted advocating for his visionary and strategic ideas in the libertarian socialist forms of libertarian municipalism and social ecology. It is, perhaps, because of his Marxist beginnings that Bookchin provides a much more robust contemporary analysis of capitalism than Klein's. His understanding of capital and his prominence both make him an excellent example of contemporary promoters of the GOD ideology.

Bookchin's ecological position begins with the (altogether correct) position that capitalism is based on constant growth. Because 'capitalism is fundamentally based on continual accumulation, it must necessarily undermine its own ecological base' (White, 2008: 84). Growth under capitalism was necessary for its continuation and, thus, could not be checked, and this would lead to environmental disaster if we allow capitalism to continue to exist. Because of this position, Bookchin even rejected class struggle as a method for ending capital, arguing that in the contemporary period, it would be cross-class alliances of humans that would change our social relationships out of a shared concern for environmental protection and human survival. With this in mind, he even eviscerated the Marxists of the late 1960s era for what he saw as their single-minded focus on class and political economy (Bookchin, 1969).

In his analysis, Bookchin discounts the possibility of different capitalisms emerging with new methods of state intervention to sufficiently respond to environmental crises and, thus, fails to consider how a dynamic capitalism could overcome such crises and co-opt radical action against it (White, 2008). His Marxist roots shine through here:

> If any serious ecological conclusion is to be drawn from *Capital Vol. I*, it is from Marx's compelling demonstration that the very law of life of capitalist competition, of the fully developed market economy, is based on the maxim, 'grow or die.' Translated into ecological terms, this clearly means that a fully developed market economy must unrelentingly exploit nature. (293)

Growth is certainly central to how capitalism operates, but Bookchin here does not account for growth, profit-making and capital accumulation as it occurs in an information economy and in a political-economic environment that, at times, rewards environmental repair with capital. In our contemporary world, where services and desires are created and exploited in virtual worlds, disconnected from the exploitation of nature (outside, of course, the harmful mining and manufacturing processes that first went into creating this digital world), it should at least give us pause to reconsider the GOD thesis. Are there, perhaps, forms of growth – particularly in a highly financialized economy – that might be disconnected from the ruinous effects that contemporary capitalism largely has on the natural world? Further, in an age of cap and trade, could growth also be achieved by *providing* environmental services rather than being built on environmental decline? Bookchin's GOD ideology prevents an analysis of those kinds of possibilities.

There are certainly more catastrophist theorists who have lit the way for environmental and political-economic thinking within radical political ecology. But Marx, Luxembourg, Klein and Bookchin allow us to engage with some major thinkers both historically and contemporarily who were/are perhaps the best-known proponents of these ideas while unpacking the nature of their arguments. In our next section, we will highlight some contemporary and historical examples of state and capital intervention into the problems we described earlier. Doing so, by pointing out these possibilities on a large scale, we aim to demonstrate that it is *possible* that capitalism might be able to create its own release valves (again, setting aside the *probability* of this happening), an alternative not explored by catastrophists. Perhaps we – or at least those of us who might emerge into Sustainlandia –are not automatically on a collision course with ourselves in a capitalist political economy. If not, that raises some interesting questions about the basis of what a coherent radical political ecology might look like. So, imagine, if you will, how some of these processes and proposals might look scaled up to rescue global capitalism from itself.

COMMODIFYING RESISTANCE

Capital and the state operate under the precondition of capital accumulation. If events transpire that threaten this accumulation, it is necessary for states to intervene.

Sometimes the state intervenes directly on behalf of specific capitalists, but there are also social moments when it has acted for the immediate benefit of labour (to rescue the capacity for capital accumulation in the long term). It is worth considering the large ways that the state has intervened in larger political and economic activity to witness its scale and to consider the possibility that states might create and support green and poverty-conscious initiatives in order to secure the continued possibility for capital accumulation.

Consider, as one example, bank deregulation (which happened alongside many state projects in the neoliberal period rooted in economic deregulation for industry) in the context of the Chrysler bailout in 1979. The 1933 Glass-Steagall Act emerged from the momentous occasion of economic crisis, where the state seemingly intervened on behalf of workers. Glass-Steagall was a historic bill that forced a separation between commercial and investment banking, regulating the power wielded by the banking industry, thus breaking up financial monopolies and decreasing the power of finance firms. But from 1933 onwards, the leaders of financial institutions collectively leaned on the state to chip away the regulations that came from Glass-Steagall to re-capture their market dominance. In the late 1970s, the Chrysler Corporation was in crisis, and American financial institutions lobbied for the erosion of Glass-Steagall as a quid-pro-quo for providing bailout funds for Chrysler. With the looming loss of Chrysler (and its concomitant jobs, investment, etc.) in the background, the banks successfully lobbied for the removal of many of Glass-Steagall's regulatory features.

These features were further eroded by the Graham-Leach-Bliley Act (GLBA) in 1999. The GLBA was to serve the ostensible purpose of 'modernizing' the banking industry to adjust for an increasingly globalized capitalism. President Bill Clinton signed the GLBA into law with the effect of formally repealing sections of Glass-Steagall that prevented single financial firms from operating as investment, insurance and commercial banks. Thus, already large financial institutions that served different sectors of the economy were allowed to consolidate and integrate, swallowing up both competitors *and* different market sectors without federal regulatory agencies having the power to intervene.

This followed a successful 1998 merger of Citicorp and Travelers Group, a merger prohibited by Glass-Steagall and, thus, a de facto challenge to its already-eroded authority. Citicorp was a commercial bank holding company while Travelers Group was an insurance company. Their merger, then, challenged Glass-Steagall's regulations by forming Citigroup, which would function as a company that dealt in securities, insurance and services under the umbrella of a variety of business monikers – combining sectors of the market that Glass-Steagall sought to split. The Federal Reserve met this challenge, not with regulatory control, but with permission in the form of a temporary waiver for Citigroup until, just months later, the GLBA was signed into law, legally allowing these kinds of market mergers of financial firms.

This is an example of the state engaging in large-scale political-economic activity in order to alter the functioning of capitalism, in part as a result of the lobbying efforts of financial institutions. Not only did the state bailout the Chrysler Corporation, but it allowed the general deregulation for the financial industry. We are still feeling the effects of these processes today. Indeed, it was this deregulation that led to the 2008 financial crisis, where the state yet again intervened on behalf of capital in the form of another bailout – this time for the banks. But the state has an uncanny way of co-opting possibilities that might, at first glance, look like forms of resistance and these large-scale, structural activities on behalf of capital might be re-tooled as efforts that benefit labour or the environment. Consider, for example, the policies of the New Deal era or the creation of social democracy in Western and Northern Europe. What if the financial and legislative efforts that went into the Chrysler bailout and deregulation were instead put into green energies and social safety nets? What if the dollars that went into the bank bailout after the mortgage crisis in 2008 were put into the hands of families struggling to keep their homes? These sorts of questions demand that we consider the possibilities of state intervention to create a green and need-aware capitalism, to co-opt the energies in contemporary environmentalism and class struggle.

Universal Basic Income

Perhaps no co-opting, capitalist political-economic programme has captured the public imagination recently like the idea of a universal basic income (UBI). The idea behind the UBI is fairly simple. Every citizen of a given country is given a livable income without conditions, no need to work, no need to look for work, and so on. Rather, this income is universal, given to all citizens (which, of course, brings into question the attainability of citizenship). The idea is an old one, but today it is often posited as a cure for increasing automation and the loss of jobs it might bring.

Right now, the experiments in UBI are fairly small, but some do reach the level of small cities. There is an experiment run by a non-profit in western Kenya that began in 2016. State experiments are ongoing in Utrecht in the Netherlands and in Ontario, Canada. Finland has instituted an experiment, including 2,000 test subjects. Since most of these experiments began in the last couple of years (2016 and 2017), we are yet to fully understand the possibilities with a UBI. However, if these smaller experiments yield positive results, it is again worth asking questions about the prospects of scaling up.

The notion of a UBI has been criticized as a possible assault on state-provided services. Acting as a replacement for basic social welfare services, including but not limited to, healthcare and public education, for example, a UBI can be seen as a strengthening of the neoliberal assault on public goods. What if the replacement costs of healthcare and public education are cheaper under a UBI? The state, then, could use it as a method of replacing basic social services. But what if the UBI is proposed in addition to public services? This could address human necessities while also providing a small bit for personal development. Thus the experiments in the Netherlands, Canada and beyond could be retrofitted to provide both social services and an income for personal development, providing a compelling alternative to neo-liberal, laissez-faire models and showcasing capital's soft hand.

Environmental Destroyers and Saviours

Few examples of global environmental crises boast as hasty a resolution as the achievements of the Montreal Protocol. In 1987, policymakers sat down in Montreal to sign what would become an internationally hailed treaty to phase out the release of chlorofluorocarbons (CFCs) that had created a dangerous hole in the ozone layer. Never mind that the gaping hole largely was their doing in the first place: CFCs were initially developed during the Second World War as refrigerants and firefighting materials used in military aviation and food storage. Almost every government in the world sat down at the negotiation table to coordinate the phase-out of CFCs from industrial production after having expelled immense quantities as the byproduct of military and commercial production in core nations. The conglomerate of states wasted little time in rallying the science, support and commitments necessary to pull off the international agreement: in 1974 an international coalition of scientists had formed to clarify the harmful effects of CFCs and, only a handful of years after, international negotiations began to hash out the effects of ozone-depleting substances, and coalesce international regulatory efforts to curb production and harm. Developing and core nations alike quickly coordinated international research less than a decade after the initial discovery of ozone depletion. The Montreal Protocol is contemporarily recognized as one of the most successful instances of international cooperation in swiftly responding to global environmental pressure (though, of course, riddled with all the familiar inequalities in responsibility and vulnerability).

Were it not for the cleanup legislated by international negotiations, the quickly eroding protective layer of the Earth might not have been recovered, never mind the countless industries threatened by increased exposure to ultraviolet light. In the face of imminent danger to health (i.e., capital), the state stepped in to curb the excesses of capitalism when social reproduction was threatened. A study, as recently as 2016, proclaimed that the ozone layer was 'healing' after decades of regulation (Solomon et al., 2016).

Of course, the most natural comparison to such a case study is the ever-deepening crisis of climate change. Climate change is perhaps the most existential threat facing humanity thus far, and the threat that mitigation and adaptation pose to capital accumulation are immense but known. A more dire situation than the threat posed by a depleting ozone layer, the issue of climate change demands similar criteria for some level of success (i.e., survival): international coordination of scientific research; adequate and fair funding for adaptation, technology transfers and disaster; and global regulatory agreements that are binding and enforceable, among many other key components of achieving climate resiliency. Decades of international scientific collaboration and increasing public clarity about the anthropogenic roots of global warming finally produced the Paris Climate Agreement, coming out of the 2015 United Nations Climate Change Conference.

The Paris Climate Agreement does point to the beginnings of international coordination to address climate change and the multidimensional threats it poses to capital reproduction and governability. In the global media spectacle that was the

UN climate conference, the treaty boasted near-full global participation in an international pinky promise to meet voluntary and non-binding targets. This is to say nothing of the market-initiated projects towards sustainable production, like shifts in energy and agricultural policy (more further here). But we do, at the very least, understand in broad strokes the breadth and timescale at which we must coordinate international climate action; the political and economic will to do so is anyone's guess. But the evidence exists to support that capitalists are paying attention, and green capitalism is on the menu. Market solutions alone might not avert catastrophe, but it's the state that makes such co-optation possible. Despite Trump's intentions to withdraw from the agreement, other countries are pushing forward with the targets set by them, putting into motion observable policy changes towards a renewably powered, climate-resilient world. Another future, aside from catastrophe, is possible, and it could come in the shape of progressive capitalist politics.

A key component to addressing the crisis of climate change is, of course, the ability to move away from fossil fuels to renewable sources of energy. There are several notable trends in the political economy of energy that give reason to believe there are various versions of capitalism that either currently exist or are foreseeable as viable routes through which ecological stress can effectively be addressed, while maintaining capital accumulation, the state and hierarchy.

Re-Energizing Capital

Energy is rapidly becoming a site of green development, and there is undeniably a notable (if disorganized) coalition of capitalists with material interests in facilitating the transition to greener forms of production. Northern European countries, for instance, have already demonstrated the ability of certain forms of capitalism to respond to energy pressures with the European Union currently boasting 16.7 per cent share of energy from renewable sources and aggressive goals for 100 per cent national renewable energy use as early as 2040 (Eurostat, 2017). Countries in the periphery have made similar but limited gains. Costa Rica has run almost entirely on renewable energy for the past two years, generating its own power from a combination of geothermal, hydro, wind and solar energy. India, a recent and particularly aggressive polluter, has some of the most rapidly expanding solar panel production in the globe, quadrupling its solar power capacity in just the past three years. India alone has an extensive land capacity for solar energy that, when at full productive capacity, has enough solar energy available in a year that exceeds the possible energy output of its entire fossil fuel energy reserves.

In the United States, the cost of solar energy is steadily decreasing and is likely to soon become competitive with more carbon-intensive forms of electricity generation. Researchers at Stanford recently released a report laying out how the United States could transition to complete renewable energy by 2050 with carefully executed government subsidies and corporate investment, which are, of course, all within the confines of the current political-economic institutions (Jacobson et al., 2017). The US Department of Energy recently reported that the solar energy industry now employs more people in US electricity generation than the oil, coal and gas industry combined. The price of residential and industrial solar energy system installations is steadily dropping across the United States (US Department of Energy, 2017).

These examples aren't without their limits and challenges, especially with respect to scale, but they do point towards the possibility of expanding such developments. Perhaps the current political-economic arrangement has the capacity to fork up the trillions of dollars of investment necessary to avert total crisis. Of course, crisis won't be averted for most: climate and other environmental disasters consistently affect low-income communities of colour the first and worst. Systems like colonialism, white supremacy, patriarchy and capitalism leave oppressed groups vulnerable to the effects of a problem they, in large part, did not cause. These (bourgeois) revolutions come with the familiar collateral damage that accompanies any process of commodification, thus the need for turning our analytical attention to Ironborough as it relates to Sustainlandia. Nevertheless, with these limits in mind, consider the possibility of these examples 'scaling up' globally. These are already large-scale, institutional examples of what is being touted as 'green' energy and, indeed, the environmental effects of these forms might help mitigate the environmental harm embedded in contemporary energy production. Imagine global subsidies for the development of green energy alternatives and international cooperation in their development. Not only could green energy be cheap, but it could also be *profitable*. That is, it is possible that we could make rescuing ourselves from environmental catastrophe lucrative.

Subsidizing the Next Green Revolution

Industrial agriculture under neoliberal capitalism tells a similar story, where one way to address environmental (and perhaps political-economic) decline in the context of agriculture is a simple moving of subsidies. This was recently the subject of discussion in Germany, where 'protesters voiced their support for directing EU agricultural subsidies to farmers who operate ecological farming practices instead of paying farmers based on the total amount of land they use' (DW, 2018). Currently, in the United States, we spend millions of dollars in subsidies to grow large-scale, industrial-size monocultural fields of corn, wheat and soy through capital-intensive agriculture (i.e., through a combination of mechanization and heavy pesticide use).

Historically, state subsidies for corn were intended to create a surplus of grains to feed the population, possibly through wartimes (when production levels might decrease) or in times of drought or other environmental crises. But the corn lobby seized on this history to ensure that the subsidies remained. Thus, we currently find things like high fructose corn syrup as a cheap sugar replacement or caloric filler in a lot of our foods. Now corn, because it is so cheap due to subsidies, is often used to feed industrially produced cattle in large-scale animal agriculture operations. Thus, the soil is depleted in the farming method, and methane emissions are increased in cattle production.

A possible fix to intervene in this large slice of agriculture (where industrial monocultures lead to decreasing biodiversity, soil degradation, increased levels of

CO_2 emissions, as well as methane emissions in the attendant animal agriculture) would be to just move those existing subsidies. Instead of spending millions of dollars every year in existing farm bills, we could move those subsidies to small-scale, labour-intensive, polycultural forms of farming. Again, this doesn't require ending capitalism. Quite the contrary, there is already a large existing niche market for green agriculture (though its 'green-ness' might be questioned). But just by moving the subsidies, we could incentivize forms of farming that reduce food miles (through small-scale localization), utilize soil protective forms of farming (polycultures) that reduce carbon emissions and in some cases use carbon capture methods, that disincentivize the use of corn as a feeder for animal agriculture, also addressing those pesky methane emissions. As you can see with this example, we've done all of this *using capitalism and its state.* If that's the case for a wicked problem like this, it begs the question of what other large 'fixes' capitalism might produce while also demonstrating the necessity of a radical political ecology that opposes capitalism on its own terms without the need for catastrophist theories or rhetorical devices.

SUSTAINING EXPLOITATION

Capitalism takes us to the point of crisis only to credit itself with bringing about solutions. Political ecology needs to critique capitalism because it is a miserable, oppressive system, not because it necessarily will lead to catastrophe or its overthrow. When techno-capitalists like Mark Zuckerberg mount a platform to tout the virtues of UBI, echoing the calls of ostensible 'anti-capitalists', something isn't right. It is entirely within the realm of possibility to live in a world where domination can be sustainable – nothing about going green per se dictates that we won't still live under the whip of our bosses, that Black and Brown folks won't bear the brunt of systemic violence, or that the state's baton won't come down on us and our resistive movements. Today's intersecting crises of environment, social domination (i.e., white supremacy, patriarchy) and a relatively unregulated market society have led to a generalized crisis of legitimacy for how our dominant institutions are organized. Resistive movements have grown under this multidimensional threat, responding to the social relations that gave rise to these crises. It's not hard to imagine a world where our rulers co-opt this energy and appease some of our greener and more social sensibilities so that we tolerate oppressive social relations. Indeed, it's not hard to imagine a green, need-aware future, but perhaps even one ran by a multicultural rainbow coalition of elites, co-opting feminist and anti-racist movement(s) while simultaneously constructing this future on the backs of femme people of colour around the world. If there were a possibility to create a softer, greener and more need-aware capitalism, this would almost certainly be created on the backs of low-income folks (mostly women) of colour largely in peripheral nations who would pay the price for transition without reaping its benefits. In fact, as radicals wishing for another world, it would be unwise not to anticipate the next social-political-economic arrangement that our rulers dream up and enforce to maintain their structural power.

Capitalism historically developed because of the state's power to codify morality and institutionalize hierarchy in all of its forms (Gelderloos, 2017). It is precisely because of the state's historical relationship to capitalism as a manager of its contradictions and excesses that a greener capitalism certainly is plausible – the state is the tool through which capitalism co-opts antagonist energies, and reforms itself to survive crises of both legitimacy and material conditions. This necessitates an anti-state analysis of state functions alongside capitalism's co-opting tendencies and integrating them into political ecology.

In anticipating and struggling for the next world, it's necessary to put forward a critique of capitalism *because it is undesirable*, not because of a belief that capitalism will somehow automatically lead us to hell. We want a different world because domination is undesirable. We want a different world because we want control over our own lives. And a contemporary radical political ecology should avoid the siren song of the emerging professional-managerial green consensus, in service of producing a green and need-aware society where we still sell our dreams for the resources necessary to live. Of course, a Sustainlandia can never exist without its respective Ironborough – any 'green' future necessarily will be built on the backs of the oppressed. And for the residents of this 'green' future, we might posit Vaneigem's question (1967), 'Who wants a world in which the guarantee that we shall not die of starvation entails the risk of dying of boredom?'

NOTES

1. For somewhat opposing views, see https://thenewinquiry.com/turn-down-for-what/ and http://deontologistics.tumblr.com/post/91953882443/so-accelerationism-whats-all-that-about

REFERENCES

Bookchin, M. (1969). Listen, Marxist! [Pamphlet]. *Anarchos*.

Campbell, S. (2014). Why Read the Junius Pamphlet. *Socialist Review* 388. http://socialistreview.org.uk/388/why-read-junius-pamphlet.

DW. 2018. Tens of Thousands March in Berlin for Greener Agricultural Industry. *DW*. http://www.dw.com/en/tens-of-thousands-march-in-berlin-for-greener-agricultural-industry/a-42240540.

Eurostat. 2017. Renewable Energy in the EU: Share of Renewables in Energy Consumption in the EU Still on the Rise to Almost 17% in 2015. *Eurostat Press Office*. http://ec.europa.eu/eurostat/documents/2995521/7905983/8-14032017-BP-EN.pdf/af8b4671-fb2a-477b-b7cf-d9a28cb8beea.

Foster, J. B. (2000). *Marx's Ecology: Materialism and Nature*. New York, NY: Monthly Review Press.

Gelderloos, P. (2016). *Worshiping Power: An Anarchist View of Early State Formation*. Chico, CA: AK Press.

Jacobson, M. Z., Delucchi, M. A., Bauer, Z. A. F., Goodman, S. C., Chapman, W. E., ... Yachanin, A. S. (2017). 100% Clean and Renewable Wind, Water, and Sunlight All-Sector Energy Roadmaps for 139 Countries of the World. *Joule* 1: 108–121.

Klein, N. (2015). *This Changes Everything: Capitalism vs. The Climate*. New York, NY: Simon and Schuster.

Kropotkin, P. (1997). *The State: Its Historic Role*. London, UK: Freedom Press.

Kropotkin, P., & Grave, J. (1916). Manifesto of the Sixteen. https://contrun.libertarian-labyrinth.org/the-manifesto-of-the-sixteen-1916/.

Luxemburg, R. (1915). *The Junius Pamphlet: The Crisis of German Social Democracy* [pamphlet]. https://www.marxists.org/archive/luxemburg/1915/junius/index.htm.

Mackay, R., & Avanessian, A. (2014). *#Accelerate: The Accelerationist Reader*. Windsor Quarry, UK: Urbanomic.

Marx, K. (1981). *Capital: A Critique of Political Economy*, Volume 3. New York, NY: Vintage Books.

Marx, K., & Engels, F. (1848). The Manifesto of the Communist Party. https://www.marxists.org/archive/marx/works/1848/communist-manifesto.

Moore, J. (2014). The End of Cheap Nature: Or How I Learned to Stop Worrying About 'The' Environment and Love the Crisis of Capitalism. In *Structures of the World Political Economy and the Future of Global Conflict and Cooperation*, edited by C. Suter and C. Chase-Dunn. New York, NY: Lit Verlag.

Moore, J. W. (2015). *Capitalism in the Web of Life: Ecology and the Accumulation of Capital*. New York: NY: Verso Books.

Sandler, B. (1994). Grow or Die: Marxist Theories of Capitalism and the Environment. *Rethinking Marxism: Journal of Economics, Culture & Society* 7(2): 38–57.

United States Department of Energy. (2017). U.S. Energy and Employment Report. https://www.energy.gov/sites/prod/files/2017/01/f34/2017%20US%20Energy%20and%20Jobs%20Report_0.pdf.

Vaneigem, R. (1967). The Revolution of Everyday Life. https://theanarchistlibrary.org/library/raoul-vaneigem-the-revolution-of-everyday-life.

White, D. (2008). *Bookchin: A Critical Appraisal*. London: Pluto Press.

Index

abstract labor, 95
accelerationism, 206
accessibility, 75
Adivasis (first dwellers), 152
AEC. *See* Atomic Energy Commission
Afghanistan, 22
AFPE. *See* Anarcha-feminist Political
    Ecology
Africa, 53, 54, 111
Agrawal, 162, 167
agriculture, 6
AIDS epidemic, 169
Aka-Bea-Da, 148
Akram, S. M., 47
alienation, 23, 134, 135
alter-globalization, 58
alternative energy, 41
altimetric satellites, 16
Amazon (corporation), 21
Amazon Basin, 54
Amnesty International, 48
Anarcha-feminism, 129, 130
Anarcha-feminist Political Ecology (AFPE),
    125, 130, 133–36
anarchic communities, 144
anarchic-gregarious form of collective living,
    71, 83
anarchic society, 146, 147

anarchic way of life, 151, 154
anarchist geographies, 3
anarchist political ecology, vii, xii, 10, 92;
    advocating for, 6; areas of, 4; explicit
    normative commitment of, viii; on green
    economy, x; modernity and, 5; resources
    management and, 69; on violence, ix
anarcho-syndicalism, 195
ancient Rome, 207
Andaman Islands, 144, 147, 151, 156
Andaman Padauk (*Pterocaopus
    dalbergioides*), 148
Andaman Trunk Road (ATR), 150, 151,
    153
Anthropocene, 4, 8, 185, 186, 190, 194
anthropocentrism, 6, 143
anthropogenic harm, 2
anthropogenic impacts on Earth, 153
anti-capitalism, 10, 162
anti-colonialism, 82
anti-corruption, 54
antifa, 171
anti-immigrant policies, 54
Arctic Indigenous population, 69
arsenic contamination, 130
ASA. *See* Authorised Syndicated Association
Asia, 111
associative life, 72, 73

www.ingramcontent.com/pod-product-compliance
Lightning Source LLC
Chambersburg PA
CBHW050641280326
41932CB00015B/2738